A BLUEPRINT FOR
GEOMETRY

Brad S. Fulton

Bill Lombard

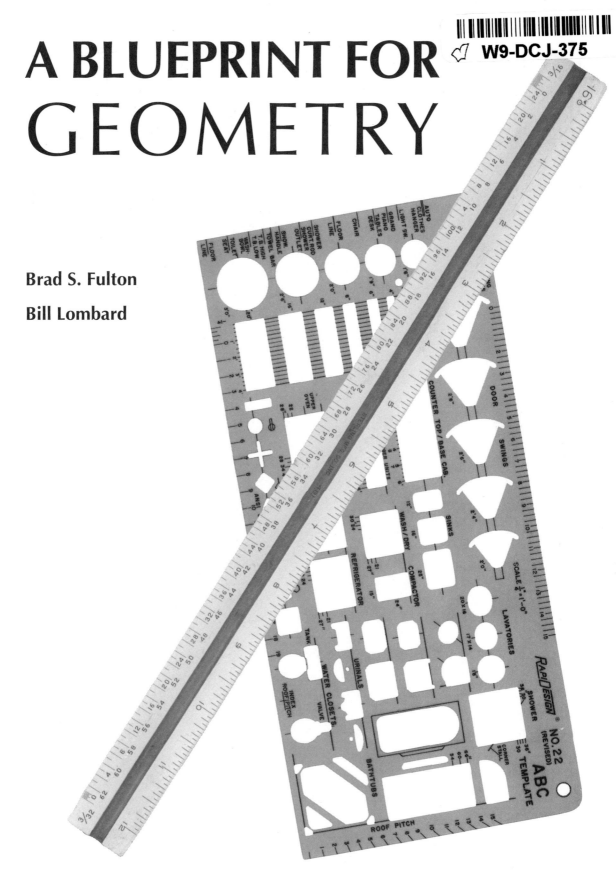

W9-DCJ-375

Dale Seymour Publications®

Project Editor: Joan Gideon
Production Coordinator: Joe Conte
Art: A. W. Kingston Publishing Services
Text and Cover Design: Christy Butterfield

Published by Dale Seymour Publications®, an imprint of Addison
Wesley Longman, Inc.

Copyright © 1998 Dale Seymour Publications®. All rights reserved.
Printed in the United States of America.

Limited reproduction permission. The publisher grants permission to
individual teachers who have purchased this book to reproduce the
blackline masters as needed for use with their own students.
Reproduction for an entire school or school district or for commercial
use is prohibited.

DS21842

ISBN 1-57232-278-0

5 6 7 8 9 10-SCG-0302 01 00 99

This book is printed on
recycled paper

CONTENTS

INTRODUCTION

A Blueprint for Geometry is a curriculum unit written by teachers for teachers and students. It is based on specific and important goals defined by the National Council of Teachers of Mathematics *Curriculum and Evaluation Standards for School Mathematics*. The Standards call for challenging investigations that develop thinking skills in solving problems within real-life contexts. Page vi shows the alignment between the mathematics of each activity and the strands of the Standards.

Middle–school students are students in transition. They learn from the concrete, but they are capable of moving into the abstract. This unit will help them make this transition. Working with the process of designing and drawing to scale a house gives students opportunities to practice and master mathematical skills in a setting that is real. They will learn about the skills needed by architects, engineers, and builders who design and create the buildings we live in. They will learn how to read blueprints, the tool of a builder, and create their own blueprints. They will learn to take measurements and reduce them to a scale that can be drawn on paper.

Mathematics Content

In completing this unit, students will use most of the concepts in a geometry chapter, particularly measurement and scale drawing. The authors have found that approaching these areas of instruction through designing a house maintains a high level of student interest within the context of real-life problem solving.

Students will use the following mathematics skills while solving the problems in this unit.

ACTIVITY	SKILL
1	making a scale model, measurement, working cooperatively
2	measurement, computing area, working with scale, working cooperatively
2A	understanding linear measurement using fractions of a unit
3	making a scale drawing, making a three-dimensional model, measurement, spatial visualization
3A	measurement
4	collecting data, finding central tendency
5	spatial visualization, three-dimensional isometric drawings
6	measurement, area, cost, problem solving, working cooperatively
6A	discrete math, measurement
6B	measurement, perimeter, estimation
6C	measurement, area, estimation
6D	measurement, volume, estimation
6E	measurement, area, ratio, percent, estimation
7	spatial visualization
7A	measurement, Pythagorean theorem
8	measurement, problem solving, geometry, area, cost, estimation, working cooperatively
8A	measurement, problem solving

In addition, journal prompts are given for each activity. These incorporate mathematical reasoning, logic, and language into each lesson. Journal prompts can be assigned in class, used as part of the homework, or used to generate class discussion.

Planning the Time

The authors are middle-school teachers currently at work in the classroom. They know that time is precious. A teacher with a full curriculum can't afford to spend six to eight weeks on a replacement unit that only covers a single topic. This unit is designed to be completed within three to four weeks. Ten extensions are included so that more time can be spent on the unit if additional time is available and students are interested.

Activities designated by a number and letter (such as 3A) are optional. They serve one of two purposes. Some are designed to provide additional instructional opportunities for students needing a skill necessary for the activity. For example, Activity 3A introduces students to measurement with fractions of an inch, a skill initially required in Activity 3. Others provide more extended instruction for students requiring a more challenging curriculum. An example of this is Activity 7A, which uses the Pythagorean theorem. Homework is provided for most of the activities, as well as ideas for extensions and investigations. Assessment ideas are suggested when appropriate. A general time guideline follows.

Activity	Periods Required	Optional Activity	Periods Required
1	1–2		
2	1		
		2A	1
3	2–3		
		3A	less than 1
4	1		
5	1–2		
6	1–2		
		6A	1
		6B	1
		6C	1
		6D	1
		6E	1
7	1–2		
		7A	1
8	2–5		
		8A	1
Total	**10–18 Days**	**Options**	**8 Days**

Classroom Visits by Professionals

If possible, invite professional architects into the classroom to discuss their work and to introduce the activities.

Their visit can be a general introduction to their own occupation or specific to the activities of the unit through introducing terms and showing real-life examples. Encourage them to use vocabulary used by architects, and show examples of their work. Ask them to bring floor plans, pictures of buildings, cost estimates, and whatever other tools architects use. This will help motivate students when they tackle the activities themselves.

Alignment

This unit strongly addresses the NCTM Standards in these areas.

Math as Problem Solving. Designing a home within a given budget, analyzing the square footage of homes.

Math as Communication. Writing in journals, conducting class discussions, working cooperatively, making class presentations.

Math Connections. Studying the connections between two-dimensional and three-dimensional models.

Computation and Estimation. Computing area, perimeter, and estimating cost.

Geometry and Spatial Sense. Working with length and area, designing two-dimensional and three dimensional models, matching elevations to floor plans.

Measurement. Measuring length, area, and volume.

Statistics. Collecting data, using charts and graphs.

MATERIALS

All of the necessary drafting materials can be created from the blackline masters. However, if available, it is best to use actual drafting tools.

Required Materials

12-inch by 18-inch white drawing paper

Copies of the large floor plans

Standard rulers

Yardsticks or measuring tapes (It is good to have an assortment of lengths. Measuring tapes over 3 feet roll up into housings.)

Grid paper and isometric dot paper

Optional Materials

Architectural rulers (Choose those with $\frac{1}{4}$-inch and $\frac{1}{2}$-inch scales on them.)

Drafting triangles (These are available in right triangles with 45° angles or with 30° and 60° angles.)

Drafting templates (Though a class set of about 10 quarter-inch drafting templates may be expensive, they last for many years and lend a professional touch to the house plans.)

A Blueprint for Geometry poster

A small amount of preparation will be necessary to begin this unit. After reading through the activities, you will have a good idea of what will be needed at each stage. You can save some prep work later by anticipating those needs before beginning the unit.

Included in the materials provided with this book are four large plans as well as 12 additional plans with corresponding elevations. To promote interest in the unit, make up the four full-size plans (A, B, C, and D) ahead of time. You will need to trim the margin off one sheet of paper before carefully aligning it and taping it to the next as in the following diagram.

Laminate these to create posters, if possible, and display them on the walls before you begin the unit. Students will enjoy looking at these and will be eager to start the unit. You may also wish to put these plans on a bulletin board with the elevations so students can try to match each floor plan with its elevations. Plan C, along with photographs of its construction, also appears on the poster *A Blueprint for Geometry.*

All these plans are drawn as they would be by an architect. Some of the symbols may be more complex than your students will use. Talk about the symbols, and decide with your class how they can be simplified on student drawings.

The number of copies of the plans you will need will depend on whether you want your students to complete the problem-solving portions of the unit in cooperative groups or as individuals. If you choose groups, you will only need one or two copies of a plan for the group. However, if working individually, students will each need a copy. Furthermore, you may decide that each group studies only one plan, but they study it in depth. This could be their "team plan." On the other hand, if you want each student to analyze each plan, every student will need a complete set of plans. This will require more copying and will increase the amount of time required to teach the unit, but it will provide more practice for students. You may wish to keep these plan portfolios in class so students don't lose important copies.

It is advisable to make a spare set of the over-sized plans. It is important to note that some copy machines do not reproduce actual size. Measure your copies and adjust the enlargement or reduction of the copies if necessary.

You may also wish to make transparencies of all the required masters in the unit. The materials list for each activity will tell you what is required. Certain student supplies will be used frequently. Keep a supply of $\frac{1}{4}$-inch grid paper and isometric dot paper on hand.

If you plan to use the journal prompts, decide how this will be managed and assessed. Will the students keep their own journals, or will they be kept in class? Will you give students feedback? How often?

Trim this margin.

Align plan.

Tape pages.

MAKING SCALE DRAWINGS

Student Outcome

Students will practice measurement and learn about scale drawing. They will learn to draw an accurate scale drawing of the floor plan and elevations of a room.

Materials

Required
- $\frac{1}{4}$ -inch grid paper or $\frac{1}{4}$ -inch dot paper (Some students find one type easier to use than the other.)
- transparency of $\frac{1}{4}$ -inch grid paper
- masking tape or chalk
- rulers

Optional
- tape measures
- transparency of Bath and Kitchen Plans

Time Required. 3 periods

Vocabulary

floor plan A view of a room or building as seen looking straight down after the roof and upper floors are removed. It illustrates the size, form, and relationships between the spaces of the room measured horizontally, as well as the thickness and construction of the walls, door and window openings, and fireplaces.

elevation A view of a wall or building as seen looking straight ahead (at the vertical surface). It illustrates the size, shape, and materials of the surfaces and the openings within them.

Procedure

1. Prior to class, make life-size diagrams of the four bathrooms and two kitchens from the plans on page 2. Use masking tape (or drafting tape for easier removal) on a floor indoors or chalk on the pavement outdoors.

2. Tell students that they will work in teams to make a scale drawing of one of the rooms. They will measure and draw the floor plan on the grid paper or dot paper.

3. Pass out grid paper or dot paper.

4. Tell students that each $\frac{1}{4}$ -inch square on the paper equals one square foot on the full-size layout. On the top of their papers, have them write: "Scale: $\frac{1}{4}$ inch on drawing = 1 foot in real-life."

5. Model a method for measuring and transferring measurements to the grid paper on the overhead and transparency. Measure one wall in feet. Draw that wall on the floor plan. Have students do the same on their papers. Since all four bathrooms are the same size, as are the two kitchens, you can draw all the walls with the class if you wish, or you may let students work as teams as soon as you feel they are ready.

6. Have students complete the scale drawing. You may have to show them how to measure to place fixtures.

7. An easy way to assess students' work is by using a transparency of the bath or kitchen plans and elevations.

The drawings on page 2, like all the architectural drawings in this book, show outside walls 6 inches wide and inside walls 4 inches wide. What have your students done to show the width of the tape in the model? If the tape were to scale, how thick would the wall be? You may want to show them the Bath and Kitchen Plans transparency and talk about the wall thickness. If you will allow your students to use a simple line for all walls, let them know.

Extension. If a team finishes early, they have room on the paper to draw the other rooms. You may also wish to have them try to draw the elevations of the room. The walls should be made 8 feet tall and should show the fixtures on that wall. Be sure to identify the fixtures, using the Architect's Key on page 6.

Journal Topic. How is your scale drawing like the life-size diagram? How is it different? In what ways are all the different bathroom plans similar?

Homework. Measure a room at home and make a $\frac{1}{4}$ -inch scale drawing of the floor plan and elevations.

BATH AND KITCHEN PLANS

Scale: $\frac{1}{4}$ inch = 1 foot

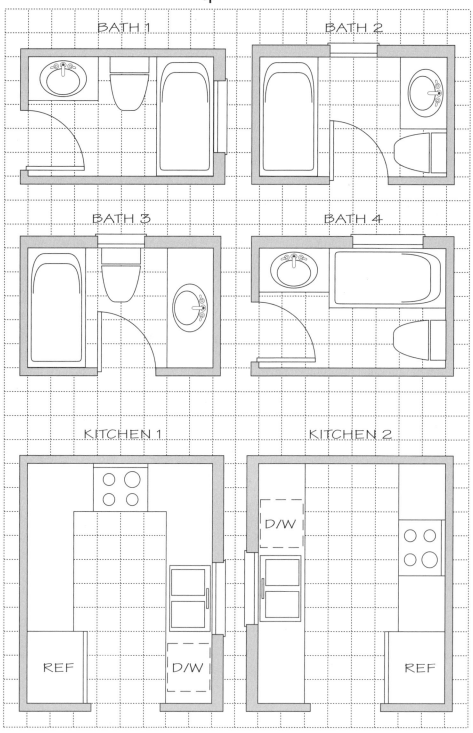

© Dale Seymour Publications®

© Dale Seymour Publications®

$\frac{1}{4}$-INCH DOT PAPER

© Dale Seymour Publications®

WORKING WITH A FLOOR PLAN

Student Outcome

Students will use a $\frac{1}{4}$-inch-scale plan of an apartment (page 8) to compute actual size. Students will learn to use an architectural ruler, if available, or to read a standard ruler and convert the marks into the appropriate scale units.

Materials

- copies of Architect's Key
- copies of Architect's Scale or Graduated Ruler
- copies of Apartment Plan, Cabin 1, and Cabin 2
- standard rulers
- transparencies of Architect's Key, apartment and cabin plans, and Architect's Scale
- tape measures (homework)

Time Required. 1 period

Procedure

1. Pass out Apartment Plan; ask students to get out standard rulers.

2. Show the Apartment Plan transparency and explain that they are going to learn how to compute the square footage of that floor plan. Show the Architect's Key transparency to introduce the symbols found on a standard floor plan. Explain that their plan doesn't include all the architectural symbols, however, they should expect to see these symbols on the plans of professionals. The architectural symbols

used on the poster, *A Blueprint for Geometry*, are printed on page 59. You can display this list as well.

3. Using the Apartment Plan transparency, talk about the conventions of measuring a plan. Tell the students they will measure from the center of one wall to the centerline of the next.

4. Show them how to use the standard ruler to measure the floor plan first. Tell them that most people will have a ruler available so they should know how to use it for this purpose. Show them that they must count the $\frac{1}{4}$-inch marks on the ruler.

5. Show the students the architect's scale by laying that transparency over top of the ruler transparency. Explain to them that architects and builders use this tool to make the translation of real-life to a small scale very easy. Ask students which tool they would prefer. They may wish to write the architectural scale below the scale on their ruler. Thus below 1, 2, and 3 inches, they would write the numbers 4, 8, and 12 to represent the real apartment's measurement in feet.

6. Give students time to measure the rooms of the apartment and fill in the worksheet. Ask them to compute the total square footage.

7. Have students compute the square footage of the other plans if time permits.

Extension. An electrical outlet is shown on one of the walls in the living room. The code for placing electrical outlets says that no place on a wall can be further than six feet from an outlet. In addition, walls less than two feet in length do not need an outlet. Draw in outlets on the walls of this apartment so that you use the fewest number of outlets required. Code requires only one outlet in the bathroom.

Journal Topic. Make a list in your journal of ways in which this plan is similar to the house or apartment where you live. List ways it is different.

Homework. Have students take home the Home Measurement Sheet to fill out for use with Activity 3 if you are not going to do Activity 2A. You may need to loan tape measures to some students.

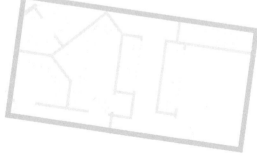

ARCHITECT'S KEY

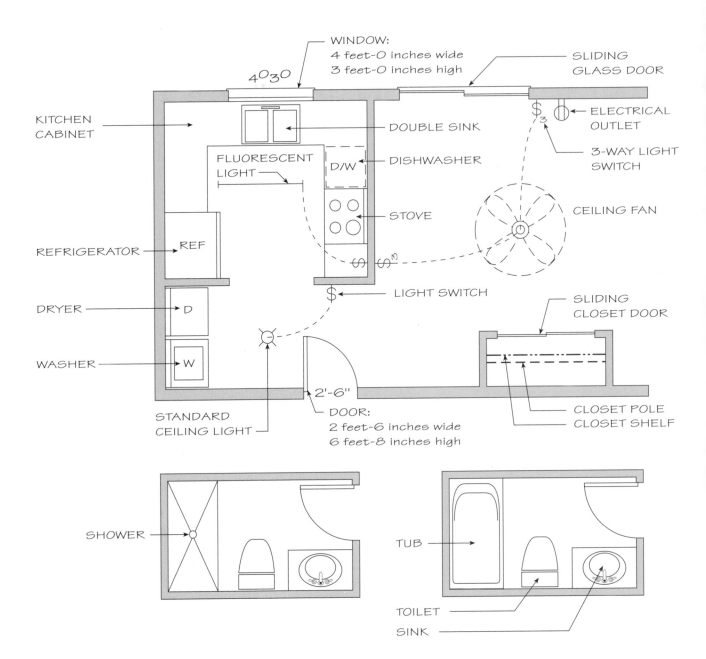

WINDOW:
4 feet-0 inches wide
3 feet-0 inches high

SLIDING GLASS DOOR

4^030

KITCHEN CABINET

DOUBLE SINK

ELECTRICAL OUTLET

3-WAY LIGHT SWITCH

FLUORESCENT LIGHT

DISHWASHER

D/W

STOVE

CEILING FAN

REFRIGERATOR

REF

LIGHT SWITCH

DRYER

D

SLIDING CLOSET DOOR

WASHER

W

STANDARD CEILING LIGHT

2'-6"

DOOR:
2 feet-6 inches wide
6 feet-8 inches high

CLOSET POLE
CLOSET SHELF

SHOWER

TUB

TOILET

SINK

© Dale Seymour Publications®

© Dale Seymour Publications®

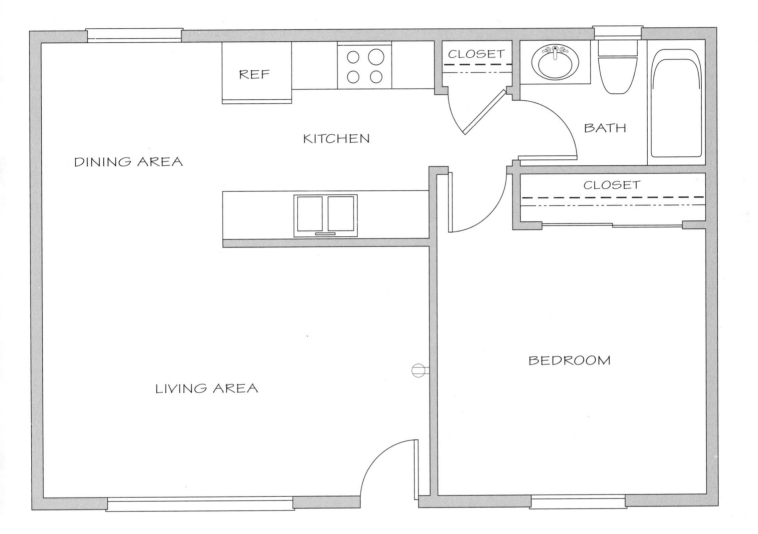

DINING AREA

REF

KITCHEN

CLOSET

BATH

CLOSET

LIVING AREA

BEDROOM

© Dale Seymour Publications®

APARTMENT SIZE

Room	Drawing Size	Actual Size
Living	_____ x _____ = _____	_____ x _____ = _____
Kitchen	_____ x _____ = _____	_____ x _____ = _____
Dining	_____ x _____ = _____	_____ x _____ = _____
Bedroom	_____ x _____ = _____	_____ x _____ = _____
Bath	_____ x _____ = _____	_____ x _____ = _____
Large closet	_____ x _____ = _____	_____ x _____ = _____
Hall closet	_____ x _____ = _____	_____ x _____ = _____
	Totals	_____

Does the sum of the areas of all the rooms equal the total area of the apartment? Why or why not?

© Dale Seymour Publications®

ACTIVITY 2A

WORKING WITH RULERS

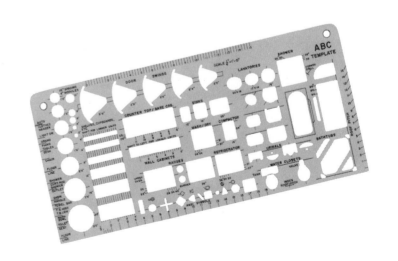

Student Outcome

Students will gain practice in using rulers with divisions of $\frac{1}{2}$ inch and $\frac{1}{4}$ inch.

Materials

Required
• graduated rulers
• Home Measurements worksheet

Optional
• $\frac{1}{4}$-inch grid paper
• tape measures

Time Required. 1 period

Procedure

Since measurement is such an integral part of this unit, this activity can be used if students are not familiar with using rulers or are having trouble measuring to the nearest $\frac{1}{4}$ inch.

1. Pass out paper rulers. Ask students to use the 1-inch ruler to measure their pencils.

2. Record the class results on the board or on a graph. You may wish the students to make their own graphs. The $\frac{1}{4}$-inch grid paper can easily be used to make a bar graph.

3. Most likely some students will report measurements such as $5\frac{1}{2}$ inches. Encourage such precision but ask them to round their measurements to the nearest inch for now.

4. Next ask them to repeat the measurement using the $\frac{1}{2}$-inch rulers, rounding to the nearest half inch.

5. Repeat step two. This will refine the distribution of answers.

6. Lastly, repeat steps one and two

with the $\frac{1}{4}$-inch rulers. As students see the graph change, they will not only appreciate the results of using more accurate divisions of measurement, but they will also become more proficient at using fractions of an inch.

Homework. Have students take home and complete the Home Measurements worksheet for use with Activity 3. You may need to loan tape measures to students.

PENCIL LENGTHS

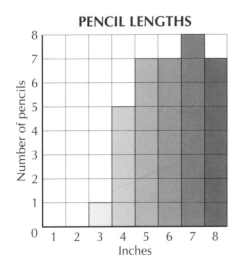

GRADUATED RULERS

© Dale Seymour Publications®

HOME MEASUREMENTS

Make the following measurements in your home to the nearest inch, $\frac{1}{2}$ inch and $\frac{1}{4}$ inch.

Measurements

Front door width	_____	_____	_____
Front door height	_____	_____	_____
Door distance from ceiling	_____	_____	_____
Bedroom door width	_____	_____	_____
Bathroom door width	_____	_____	_____
Living room window width	_____	_____	_____
Living room window height	_____	_____	_____
Window distance from ceiling	_____	_____	_____
Kitchen counter depth	_____	_____	_____
Kitchen counter height	_____	_____	_____
Kitchen sink width	_____	_____	_____
Stove width	_____	_____	_____
Refrigerator width	_____	_____	_____
Refrigerator height	_____	_____	_____
Bathtub or shower length	_____	_____	_____
Bathtub or shower width	_____	_____	_____
Toilet length	_____	_____	_____
Toilet width	_____	_____	_____
Bathroom sink width	_____	_____	_____
Bathroom sink depth	_____	_____	_____
Bathroom counter width	_____	_____	_____
Bedroom window width	_____	_____	_____
Bedroom window height	_____	_____	_____
Bedroom closet width	_____	_____	_____
Bedroom closet depth	_____	_____	_____
Hallway width	_____	_____	_____
Ceiling height	_____	_____	_____

© Dale Seymour Publications®

COMMON HOUSE MEASUREMENTS

Ceiling height	8 feet for standard ceiling
Front door*	3 feet wide, about 7 feet tall
Bedroom door*	2 feet 6 inches to 2 feet 8 inches wide, about 7 feet tall
Bathroom door*	2 feet to 2 feet 4 inches wide, about 7 feet tall
Closet depth	2 feet
Hallway width*	3 feet minimum
Counter depth	2 feet
Tub	2 feet 6 inches wide, 5 feet long
Space for toilet	2 feet 6 inches side to side
Dishwasher width	2 feet
Kitchen sink width	3 feet
Stove width	2 feet 6 inches
Refrigerator width	3 feet for standard
Washer and dryer width	5 feet total
Washer and dryer depth	2 feet 6 inches

*For wheelchair access, all doors and hallways should be 3 feet 8 inches wide.

© Dale Seymour Publications®

3-D CONSTRUCTION

Student Outcome

Students will make a scale drawing of the floor plan of a house and build a scale model of the house without a ceiling or roof.

Materials

Required

- blank paper
- Apartment Layout transparency
- Apartment Layout
- Common Household Measurements
- $\frac{1}{4}$-inch paper rulers
- scissors
- tape or glue
- Corner triangles (blackline master on page 18 copied onto card stock)

Optional

- $\frac{1}{4}$-inch grid paper (This is helpful for students who find blank paper too hard to use.)
- drafting triangles (architectural)
- poster board for models (Standard copy paper will work adequately if necessary, but some students may want to provide their own stronger material.)
- Floor Plans A–D

Time Required. 2–3 periods

Procedure

1. If you choose to use the large Floor Plans, put them up in an accessible place for the students to study. Ask students to get out their Home Measurement worksheets, which should have been completed as homework prior to this activity and the Common Household Measurements. Pass out rulers, triangles, and paper. (Blank paper should be used if students are able to work with it, but some students may require the $\frac{1}{4}$-inch grid paper.) You may wish to pass out one copy of the Corner Triangles to each team of four students and have them cut out their own individual triangles.

2. Put the Apartment Layout transparency on the overhead projector.

3. Explain to students that they will ultimately make a $\frac{1}{4}$-inch scale model of the apartment. They will begin by drawing an accurate floor plan, upon which they can construct the model. Once they have finished the floor plan, they will draw the elevations of each wall without showing the cabinets or fixtures.

4. Use a blank transparency sheet to show students how to start with a corner of the floor plan and draw the large rectangle of the exterior walls. Have them position the floor plan on the sheet so that there will be room around the edges to draw the wall elevations. Show them how to use the triangles to create 90° corners.

Tell them that drawings need to be accurate since builders take their measurements from the drawing as they build. Students should consider the thickness of their pencil point and the angle they are holding the pencil at as they measure and draw their lines.

5. Check to see that each student has constructed this rectangle correctly. You can use a transparency of the plan to lay on top of their drawing as a checking tool.

6. Have students finish drawing the floor plan. This will probably take the rest of the period. It may be finished for homework.

7. Once students have finished their floor plans, they should proceed to draw the elevations of the exterior walls. Remind them that the ceiling height is 8 feet in this apartment and that the height of standard windows and doors from the floor is approximately 7 feet.

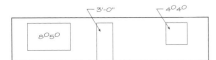

8. When students have finished with their floor plans and elevations, they are ready to begin on their models. It is best if the drawings can be glued to posterboard or cardboard to create durable models. Once the drawings are transferred to the cardboard, they can be cut out. Doors may be "hinged" so that they will open and close. Use a ballpoint pen to press hard along the line that represents the hinge side. This helps the cardboard fold neatly.

9. Students should match up their walls with the floor plan and then tape or glue them along the bottom edges and along the corners. Show them how to do this neatly by wiping off the excess glue with a damp tissue.

10. Some students with extra time will be motivated to build the counters and fixtures.

Journal Topic. How tall would you be if you were scaled to fit this apartment? Make a scaled drawing of yourself. Use this to create a cardboard person to place in your model.

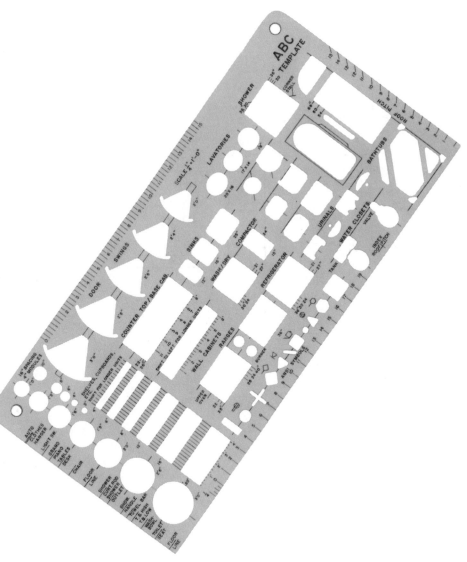

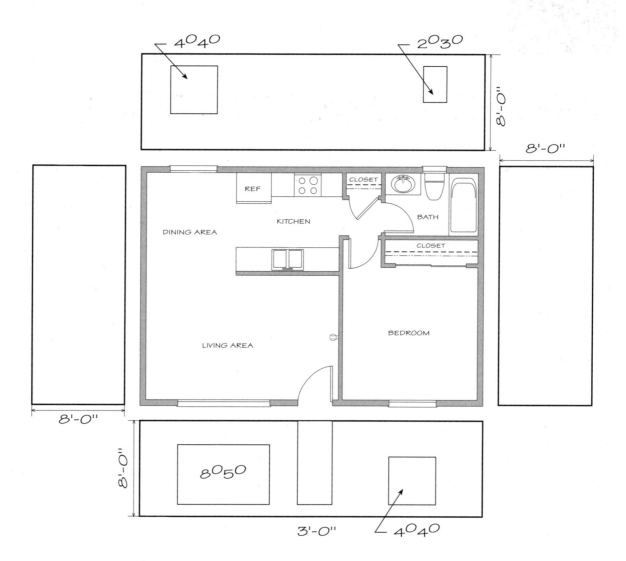

© Dale Seymour Publications®

ACTIVITY 3A

MAKING A DRAFTING TOOL

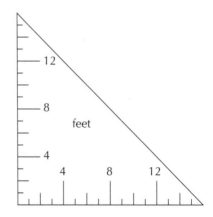

Student Outcomes

Students will construct a drafting tool (a 90° by 45° by 45° triangle marked with a $\frac{1}{4}$-inch scale on the equal legs) to use on future drawings.

Materials

- paper rulers
- Corner Triangles

Time Required. less than one period

Procedure

1. Sometimes students will invent this tool on their own. If so, use that as an opportunity to introduce it to the entire class. If you prefer, this activity can be delayed, but it should be completed before activity 8 when they draw their house plans.

2. Ask students to get out their corner triangles and their rulers.

3. Ask them to transfer their markings from the ruler to the legs of the triangle as shown. Make sure they start numbering from the square corner.

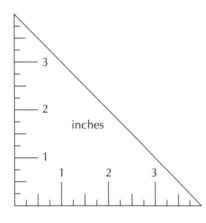

4. This tool will allow students to construct 90° corners, draw 45° angle lines, and measure lengths in inches with a single tool on future drawings.

5. Sometimes a student will take this process one step farther. The tool can be marked like an architect's scale using $\frac{1}{4}$-inch increments to represent 1 foot. You may wish to have your students mark the tool in this fashion.

© Dale Seymour Publications®

MEDIAN COST PER SQUARE FOOT

Student Outcome

Students will practice multiplication skills while learning about how to find the median cost of local houses.

Materials

- real estate magazines or real estate section of newspaper one copy per two students (You may wish to have students bring these in ahead of time for extra credit.) Be sure to find ones that have the cost of the house, the amount of square feet, and the lot size.
- Real Estate Ad blackline master—use this instead of real-life ads if necessary
- Median Cost worksheet
- Comparing Median Cost homework

Time Required. 1 period

Vocabulary

median In a set of data, the datum that is in the middle when numbers are in order. Thus half the data values are lower than the median, and half are higher. The median home price is usually more indicative of typical local housing than the mean average. In the following example, the median home price is $128,500:

$115,000
$115,500
$128,500
$165,000
$279,000

In the next example, the median ($122,000) is the average of the two middle data:

$105,500
$115,000
$115,500
$128,500
$165,000
$279,000

Procedure

1. Pass out real estate samples or copies of the Real Estate Ad sheet. Tell students why professionals use the cost per square foot. Since construction costs vary from area to area, cost per square foot is used to find what it will cost to build a home in an area.

2. Have students work in pairs.

3. Find an advertisement for a three bedroom, two bath home. It should be on a lot of $\frac{1}{4}$ acre or less. It should also show the square feet. Write its cost under the cost heading.

4. Write its area in square feet under the square feet heading on line one.

5. Divide the cost of a house by its square feet. Write that under the cost per square foot heading.

6. Have students record the data for at least seven houses (there is room for more) and find the median. If they record an odd number of house costs, the median is the one with the middle cost. If they have an even

number of homes listed, the median is the mean of the two costs closest to the middle.

7. Record medians of each team on the board.

8. Have the class find median cost per square foot based on the data on the board.

9. Use this figure on all future activities (such as Activity 8) where cost per square foot is required.

Extension. You can use the data collected by the students to introduce or practice mode, mean, and range. You can also use this data to practice making graphs. For example, give the students these five home prices. Ask them to find the mean and the median. Which is a more realistically the typical price of these homes?

$135,000
$3,375,000
$99,950
$126,900
$119,550

Journal Topic. What range of prices did you encounter in your study? What factors might have contributed to this difference?

Homework. Complete worksheet entitled Comparing Median Cost Per Square Foot.

FOR SALE: $115,900 Three bedroom, two bath house. 1/4 acre. 1,800 square feet.	**FOR SALE:** $105,500 Three bedroom, two bath house. 1/4 acre. 1,550 square feet.
FOR SALE: $135,000 Three bedroom, one bath house. 1/4 acre. 1,780 square feet.	**FOR SALE:** $117,900 Three bedroom, two bath house. 1/4 acre. 1,990 square feet.
FOR SALE: $137,500 Four bedroom, two bath house. 1/4 acre. 1,910 square feet.	**FOR SALE:** $215,500 Four bedroom, two bath house. 4 acres. 1,890 square feet.
FOR SALE: $125,990 Three bedroom, two bath house. 0.2 acre. 2,000 square feet.	**FOR SALE:** $124,000 Three bedroom, two bath house. 0.25 acre. 2,100 square feet.
FOR SALE: $105,750 Three bedroom, two bath house. 1/3 acre. 1,900 square feet.	**FOR SALE:** $119,650 Three bedroom, 1 1/2 bath house. 1/4 acre. 1,250 square feet.
FOR SALE: $135,250 Three bedroom, two bath house. 1/5 acre. 1,830 square feet.	**FOR SALE:** $159,999 Three bedroom, two bath house. 0.3 acre. 1,600 square feet.
FOR SALE: $149,000 Two bedroom, two bath house. 1/4 acre. 1,525 square feet.	**FOR SALE:** $175,500 Three bedroom, two bath house. 1/4 acre. 1,350 square feet.

© Dale Seymour Publications®

NAME

MEDIAN COST PER SQUARE FOOT

House	Cost	Square Feet	Cost per Square Foot
1.	_____	_____	_____
2.	_____	_____	_____
3.	_____	_____	_____
4.	_____	_____	_____
5.	_____	_____	_____
6.	_____	_____	_____
7.	_____	_____	_____
8.	_____	_____	_____
9.	_____	_____	_____
10.	_____	_____	_____
11.	_____	_____	_____
12.	_____	_____	_____

Median cost per square foot _____

© Dale Seymour Publications®

COMPARING MEDIAN COST PER SQUARE FOOT

Here are seven sample houses in two different cities. Find the cost per square foot of each house, rounded to the nearest cent. Find the median cost for each city. Then answer the questions.

San Lombardo

House	Cost	Square Feet	Cost per Square Foot
1.	$176,500	2,450	$72.04
2.	$108,675	1,725	_____
3.	$112,365	1,870	_____
4.	$143,950	2,200	_____
5.	$106,900	1,452	_____
6.	$154,590	2,160	_____
7.	$151,990	2,055	_____

Median cost per square foot _____

Fultonburg

House	Cost	Square Feet	Cost per Square Foot
8.	$158,900	2,300	_____
9.	$158,695	1,925	_____
10.	$110,995	1,970	_____
11.	$121,490	2,090	_____
12.	$124,900	1,645	_____
13.	$157,990	2,430	_____
14.	$122,975	1,825	_____

Median cost per square foot _____

Which city has a lower median cost per square foot?

What is the difference between the two medians?

Which house is the least expensive per square foot?

© Dale Seymour Publications®

ISOMETRIC DRAWINGS

Student Outcome

Students will make isometric drawings of cubes, cube constructions, and the apartment.

Materials

- transparency of isometric cube
- cubes
- isometric grid paper
- transparencies of isometric paper
- model apartments from Activity 3
- rulers (to be used as a straightedge)

Time Required. 1–2 periods

Vocabulary

isometric drawing A view as seen from above that represents the three dimensions of the space. The vertical elements remain vertical; horizontal elements are drawn at a 30° angle from the horizontal.

Procedure

1. Start by asking students how to show someone what a three-dimensional cube looks like in a two-dimensional drawing. Tell them that true representation requires knowledge of perspective. There is a quick method to represent objects that architects use called isometric drawing. Explain that to be successful, students require only a 60° triangular grid. Show them the transparency of the isometric drawing.

2. Pass out isometric grid paper and cubes.

3. Have students place the cube in front of them so they see an edge view.

4. Have students draw this view on the isometric paper as you model it on the overhead projector. Begin by drawing the vertical edge nearest the viewer (student). This will be difficult for many students. Problems commonly occur when they try to draw the cube's face view or when they hold the paper vertically instead of horizontally.

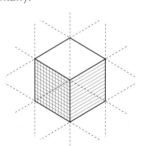

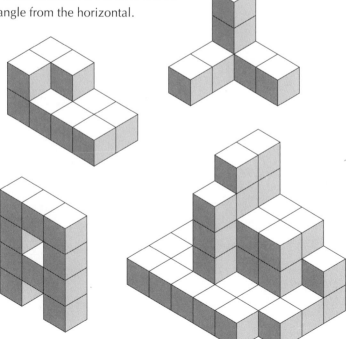

5. When students have successfully completed step 4, construct a model with three or four cubes and have students sketch it.

6. When you are confident that students understand how to use the isometric drawing grid, ask them to draw the apartment model using this method. Tell them to leave off the ceiling and roof, so that they are looking down into the apartment from the view that the isometric drawing process permits. Tell them to draw all the lines of one wall in, then erase those that are really behind a wall and shouldn't be seen.

Note: To make the drawing fit on one page of the isometric dot paper, tell students to make each line segment represent 2 feet.

Journal Topic. How many cubes would fill the apartment if each cube were 2 feet by 2 feet by 2 feet? How did you figure this out?

Homework. Draw a three-dimensional image of your initials in block lettering. How many cubes would it take to build it?

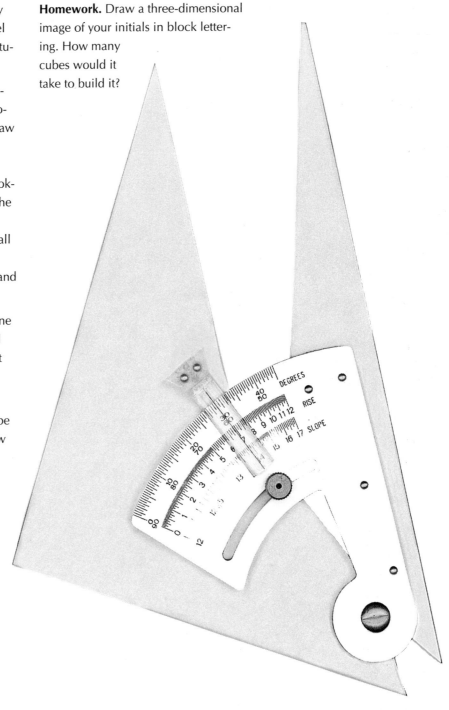

© Dale Seymour Publications®

BUILDING INSPECTOR

Student Outcome

Students will act as building inspectors and review the floor plans for accuracy and practicality. Students will compute square footage of a house plan and compute the cost per square foot.

Materials

- Plan A
- Cabin 1
- rulers
- Building Codes
- Inspection Record, two copies per student
- Common Household Measurements

Time Required. 1–2 periods

Vocabulary

building code Established by communities to ensure public safety in construction. Building codes provide safe, sound, and sanitary buildings for people to live in.

Procedure

1. Pass out Plan A, rulers, Building Codes, and Inspection Record. Explain that building codes are established for the safety of the occupants of a building and that most of the codes on the handout represent commonly accepted ones. Point out that building plans must be approved as meeting the code before a permit will be granted. A contractor must have a permit to begin construction. Inform them that this particular inspection sheet has been tailored to this small problem, but that in real life, all parts of a building must be reviewed and approved.

Tell students that the presentation of the plans is extremely important. The inspector must be able to read the plans, understand them, and consider them professional before they will be approved.

Review the Inspection Record with students to be sure they understand each item. Anything that needs attention and improvement in the plan must be reported. Explain how you will use the Inspection Record to evaluate plans students draw.

2. Have students complete the Inspection Record for Plan A. They are to use the cost per square foot from Activity 4 or a cost supplied by the teacher.

3. Have students share what they found about the plans they reviewed. Ask if they would grant a building permit. Why or why not?

Note: Plan A has livability errors.
- The window in bedroom is too narrow (building code not followed).
- There is no entrance to bedroom.
- Folding doors in master bath won't close.
- One bedroom has no closet.

Journal Topic. As the inspector at the building department, write a note telling the architect what errors you found in the plan and how to correct them.

Homework. Find the square footage of Cabin 1 and complete an Inspection Record for it.

BUILDING CODES

No bathrooms may open to the kitchen.

Exterior doors must be at least 3 feet wide.

There must be at least two entrances to the house.

Interior doors should be at least 2 feet 6 inches wide.

The toilet must have 2 feet 6 inches of clear space side to side.

Bedrooms must have at least one window 3 feet by 4 feet or larger as a fire escape.

All living areas must have a window. (Bathrooms, halls, closets, and garages are not living areas.)

Closets must be at least 2 feet front to rear.

No spot on an interior wall may be farther than 6 feet from an electrical outlet, and any wall at least 2 feet in length needs an electrical outlet. (Thus outlets should never be more than 12 feet apart.)

© Dale Seymour Publications®

INSPECTION RECORD

Architect _____

Building Inspector _____

Mathematical Accuracy   40%

☐ All measurements are correct.

☐ Square footage and cost are computed accurately.

Total area		Cost per square foot		Total cost of construction
_____	x	_____	=	_____

Architectural Neatness   40%

☐ The drawing is neat, tidy, centered, and not crowded, and writing is legible.

☐ Lines are parallel or perpendicular as appropriate.

☐ All dimensions of doors, windows, and living areas are shown.

☐ Outlets, lights, switches, and fixtures are correctly drawn.

☐ Labels are correctly oriented and spelled.

Livability   20%

☐ Building codes are followed.

☐ No extra or dead-end hallways.

☐ Traffic flow is reasonable.

☐ Rooms, doors, windows, closets, and counters are the appropriate sizes.

Score

Mathematical Accuracy (out of 40) _____

Architectural Neatness (out of 40) _____

Livability (out of 20)   _____

Total   _____

© Dale Seymour Publications®

ELECTRICAL CONTRACTOR

Student Outcome

Students will plan the number of electrical outlets needed for a house. Students will also plan the lights and switches needed.

Materials

- plans for Cabin 1 and Cabin 2
- transparencies of Cabin 1 and Cabin 2 (or transparency of page 31)
- Architect's Key or Apartment Plan
- rulers

Time Required. 1 class period

Vocabulary

electrical outlet Wall fixture into which electrical plugs are placed to get electricity; also called a *receptacle.*

Procedure

1. Pass out the plan for Cabin 1. Tell students that no electrical outlets, lights, or switches are shown on this plan. They must decide where they go.

2. First of all, ask them to show where switches and lights should go. Use these symbols.

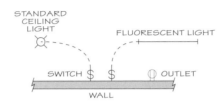

3. Remind them that the dotted lines show which switch operates each light.

4. Here are the requirements for overhead lighting.

a. Overhead lights should be in all rooms except the living room. The living room may have an overhead light if desired.

b. An overhead light should be in all hallways, stairways, and porches. Long hallways need switches at each end, and they need to be three-way switches ($_3$), so that the hall light can be turned on or off at either end of the hall.

c. The kitchen should have a light near the sink and one near the stove.

d. All overhead lights need to have a switch. The switch should be near the room's entrance (on the wall near the doorknob if there is a door).

5. Model this on the overhead projector for one or two rooms using the Architect's Key or Apartment Plan as your guide.

6. Ask students to compare plans with other members of their group to self-assess their work. You may also wish to allow some students to demonstrate their work using the overhead transparency projected onto the board.

7. Next, they must decide where to put electrical outlets.

8. The code for placing electrical outlets says that no place on a wall can be farther than six feet from an outlet. In addition, walls less than 2 feet in length do not need an outlet. Thus in this 12-foot-by-12-foot room, three outlets are needed. (No outlet is needed behind the door.)

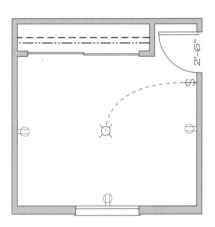

9. Remind students that not only do outlets cost money, but so does the labor of having them installed. For this reason, they should try to minimize the number of outlets; however, they should be sure that they have considered all the needs of the occupants. For instance, consider what furniture might be placed against the wall, and anticipate the ease of access to the outlets. Where do you plug in the vacuum cleaner? Do you have to move furniture to do this?

10. Using the transparency and the key as your guide, work with them in placing outlets in some rooms.

11. Have them finish placing all the outlets in their plan.

12. Next have them make sure they have outlets behind the refrigerator, washer, and dryer. The dryer requires a special 220-volt outlet. Its symbol is

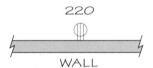

13. Let them assess their work by groups and by using the transparency as in step 6 above. You may also see who was able to use the fewest outlets. Students should use discernment in following the rule stated in step 8. Doors, fireplaces, sliding glass doors, and other obstacles will restrict outlet placement.

Homework. Have students place switches, overhead lights, and outlets on Cabin 2.

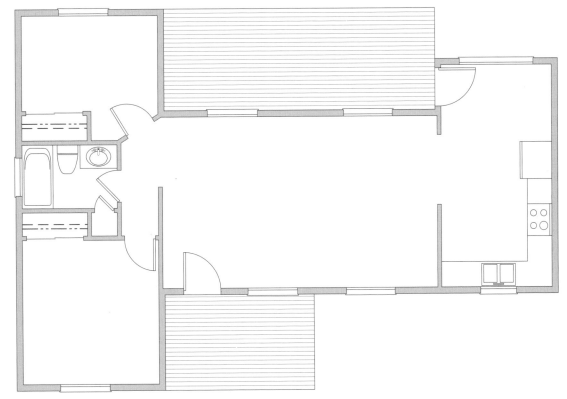

© Dale Seymour Publications®

ACTIVITY

6B

LUMBER ESTIMATE

Student Outcome

Students will estimate the number of 2 by 4 studs needed to build the walls of a home. They will determine the total linear feet of wall in the home to compute the lumber needed.

Materials

- plans for Cabin 1 and Cabin 2
- transparency of Cabin 1 and Cabin 2
- transparency of Lumber Estimator
- rulers

Time Required. 1 class period

Vocabulary

2 by 4 Standard size board. The actual size is $1\frac{1}{2}$ inches by $3\frac{1}{2}$ inches by $92\frac{1}{4}$ inches because lumber is measured to nominal size (2 inches by 4 inches) before being seasoned and planed.

linear foot One foot measurement of an item along a line.

Procedure

1. Tell students that a contractor needs to know the cost of building a house beforehand in order to know how much to charge the buyer. A part of this process involves making an accurate estimate of the lumber needed to build a house. When building with wood frame construction, the most commonly used boards, called *two by fours,* are approximately

eight feet long. When these are used to build walls, they are called *studs.* Studs are typically placed 16 inches apart along the length of a wall. However, extra studs are used at doors, corners, and windows for strength and for finishing carpentry work. A contractor estimates that when using 16-inch spacing, an average wall requires one stud per linear foot. Thus this 20-foot wall requires about twenty studs.

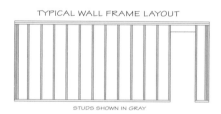

TYPICAL WALL FRAME LAYOUT

STUDS SHOWN IN GRAY

2. To estimate the number of studs required in the walls of a house, the contractor measures all the walls. The number of linear feet of walls equals the approximate number of studs needed. Thus this 12-foot-by-12-foot room with its closet walls requires approximately 58 studs.

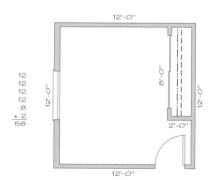

3. Have students estimate the number of two-by-fours needed to build the walls of Cabin 1. Have them show the method they used to estimate the number of studs.

Extensions

- Ask students what other parts of a house are built with lumber. Show them flooring joists and roof rafters. Have student bring in pictures of different types of roof rafters and the roofs they support. Local lumber or building supply companies are good sources for diagrams and pictures.

- Students may want to diagram the layouts of other walls in Cabin 1 or 2. Have students bring in examples of wall-framing details, including what happens at corners, around doors and windows, and how walls are joined to the floors and roof.

Homework. Have students estimate the number of studs required to build Cabin 2.

LUMBER ESTIMATOR

TYPICAL WALL FRAME LAYOUT

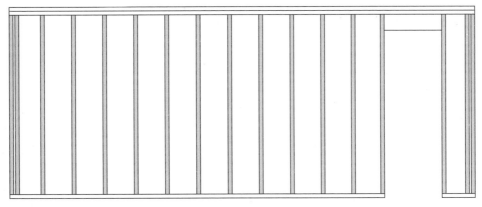

STUDS SHOWN IN GRAY

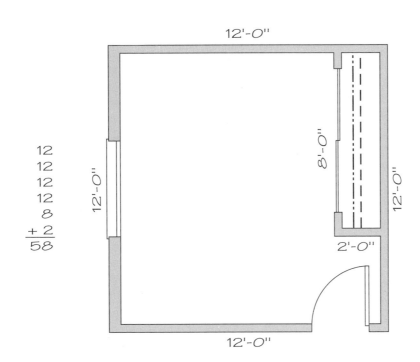

© Dale Seymour Publications®

FLOORING ESTIMATE

Student Outcome

Students will estimate the cost of floor covering needed in a house by finding the area and multiplying by the cost per square yard of flooring.

Materials

Required

- plans for Cabin 1 and Cabin 2
- transparency of Cabin 1 and Cabin 2
- transparency of Bath and Bedroom Plans
- Flooring Cost Estimate worksheet
- rulers

Optional

- Plan 1

Time Required. 1 class period

Procedure

1. Tell students that a contractor needs to know the cost of building a house beforehand in order to know how much to charge the buyer. A part of this process involves making an accurate estimate of the floor covering needed in a house. Flooring may consist of vinyl, tile, carpeting, or hardwood flooring. Show students the transparency of Cabin 1.

2. The bathroom is 5 feet by 8 feet. However, the area of the floor that needs covering is less than 40 square feet since no flooring is required below the tub. Have students discuss which flooring will be best for a

bathroom and how to estimate the amount of flooring needed. Remind them that the flooring will stop at the edge of the tub and at the edge of the vanity. Write the amount of flooring required on the bathroom of Cabin 1.

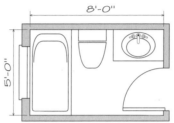

3. The bedroom on the Bath and Bedroom Plans transparency will get carpeting selling for $26.50 per square yard. Since it is 12 feet by 12 feet, it is 144 ÷ 9 = 16 square yards. (This can be figured more easily by noticing that the room is 4 yards long by 4 yards wide.)

4. Thus it will cost 16 x $26.50 = $424.00 to carpet this bedroom.

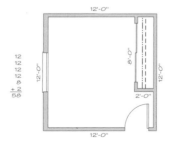

5. Have students measure the rooms of Cabin 1 and compute the total cost of carpet and vinyl. Remind them that the kitchen part of the main room in

Cabin 1 will require vinyl. Students can check their results with team members.

6. After they finish, have them help you show the solution by telling you the measurements and costs of flooring in individual rooms as you call them out from the transparency of the plan.

Extensions

- You may wish to have students compute the flooring costs for Plan 1.
- Discuss with students the different ways that flooring can be purchased.

  Vinyl flooring is sold by the linear foot and is manufactured in rolls 6 feet or 12 feet wide. Carpeting is sold by the square yard and is manufactured in rolls 12 feet wide. Tile flooring is generally sold by the square foot or in 9-inch-square or 6-inch-square tiles. Hardwood flooring can come in parquet blocks or in planks that vary in width and length.

  If possible, bring in ads from your local paper advertising different kinds of flooring.

- How many sheets of plywood (4 feet by 8 feet) would be needed if the floor of Cabin 1 were made of plywood?

Homework. Have students compute the flooring costs for Cabin 2.

BATH AND BEDROOM PLANS

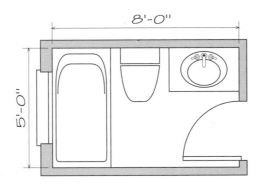

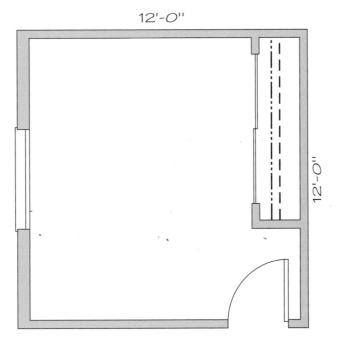

FLOORING COST ESTIMATE

FOR _____

Room	Measurements	Area	Material	Cost per Unit	Total Cost
Living area	_____ x _____	_____	_____	_____	_____
Kitchen	_____ x _____	_____	_____	_____	_____
Bathroom	_____ x _____	_____	_____	_____	_____
Bedroom 1	_____ x _____	_____	_____	_____	_____
Bedroom 2	_____ x _____	_____	_____	_____	_____
Hallway	_____ x _____	_____	_____	_____	_____

Total cost of all flooring _____

FLOORING COST ESTIMATE

FOR _____

Room	Measurements	Area	Material	Cost per Unit	Total Cost
Living area	_____ x _____	_____	_____	_____	_____
Kitchen	_____ x _____	_____	_____	_____	_____
Bathroom	_____ x _____	_____	_____	_____	_____
Bedroom 1	_____ x _____	_____	_____	_____	_____
Bedroom 2	_____ x _____	_____	_____	_____	_____
Hallway	_____ x _____	_____	_____	_____	_____

Total cost of all flooring _____

© Dale Seymour Publications®

ENERGY CONTRACTOR

Student Outcome

Students will estimate the volume of a house and select a heating system for it. This activity can be taught as an interdisciplinary unit with a science teacher.

Materials

- Familiarize yourself with the different kinds of heating systems common to your community. Bring in examples of ads for each of the systems you intend to introduce as options.
- Plan A
- rulers
- transparency of Volume of a Shed
- building cubes (snap cubes, wooden cubes, or the equivalent)
- Heating System Chart
- Cabin 1 and Cabin 2 (or another house plan)

Time Required. 1 class period

Vocabulary

forced air heating A heating system that circulates warm air through ductwork.

furnace A heater that burns coal, oil, or wood to make heat.

heat pump A heating engine that transfers heat from one area to another.

radiant heat A type of heat that uses radiation instead of forced air to heat a house.

Procedure

1. Tell students to select a heating system for a house. They need to know the volume of the air that will be heated. After discussing the merits of different systems, tell them that they are going to use a forced-air HVAC (heat vent air conditioning) system for Plan A and they need to determine which size system to select.

2. Build a model of the volume of a shed out of cubes in front of the class. Tell them the shed should be 5 feet wide by 8 feet long by 6 feet high. Each cube represents one cubic foot. Tell them to keep track of the number of cubes used to build it.

3. Ask them the square footage of the shed floor.
(5 feet x 8 feet = 40 square feet)

4. Ask them how many cubes it takes to build it (240). Tell them this is written as 240 ft^3 and read as "240 cubic feet."

5. Ask them how they could arrive at that number without building it.
(5 feet x 8 feet x 6 feet)

6. Display the Volume of a Shed transparency. Show them the four drawings, one at a time. They should see that the volume can be calculated by multiplying the length, width, and height. They should also see that they can multiply the area of the base (length x width) times the height. This latter method will work best on the house plans where the base is not a single rectangle. (It will also make future explorations of prism volumes more sensible, since the volume of a prism or cylinder is the area of the base multiplied by its height.)

7. Now ask students to find the square footage of Plan A. Tell them that typical walls in a house are 8 feet high. Ask them to calculate the volume of Plan A. (about 9,700 ft^3)

8. Ask them to use the heating system chart to select the appropriate size system for Plan A. (system 4)

Homework. Have students compute the volume and select a heating system for the other plans. (The approximate volumes for the other plans are given on page 46.)

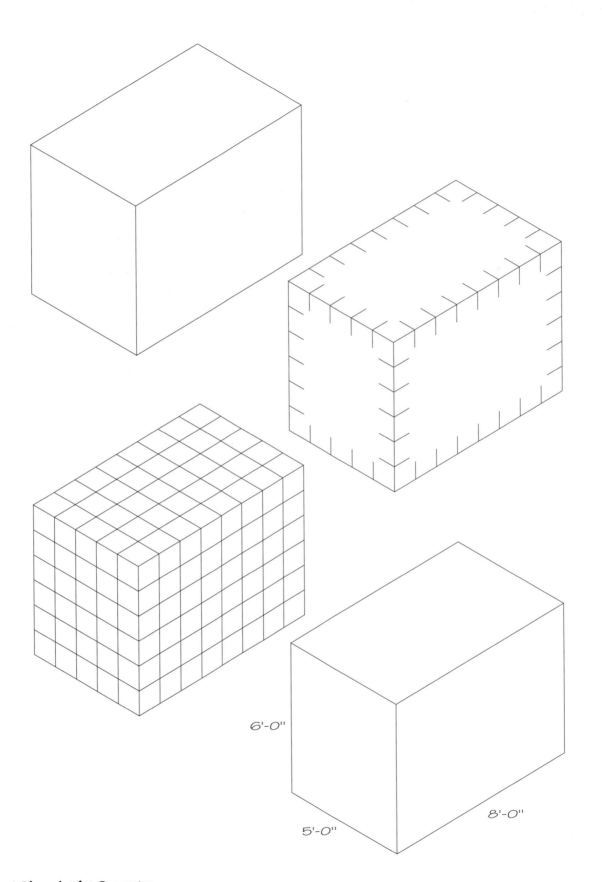

6'-0"

5'-0"

8'-0"

© Dale Seymour Publications®

HEATING SYSTEM CHART

	System	Volume Range
Electric Baseboard model B900	1	6,250 ft^3 to 7,500 ft^3
Floor Radiant Heat Panels model R1000	2	7,350 ft^3 to 8,750 ft^3
Floor Radiant Heat Panels model R1100	3	8,500 ft^3 to 9,800 ft^3
Heat Pump model P1300	4	9,450 ft^3 to 10,000 ft^3
Heat Pump model P1400	5	9,950 ft^3 to 12,500 ft^3
Furnace model F1500	6	11,750 ft^3 to 14,500 ft^3
Forced Air Heater model FA1800	7	14,050 ft^3 to 16,750 ft^3
Forced Air Heater model FA2000	8	15,750 ft^3 to 18,000 ft^3

Plan Number	Volume	Heating System
Cabin 1	_____	_____
Cabin 2	_____	_____
1	_____	_____
2	_____	_____
3	_____	_____
4	_____	_____
5	_____	_____
6	_____	_____
7	_____	_____
8	_____	_____
9	_____	_____
10	_____	_____

© Dale Seymour Publications®

6E

GLAZING CONTRACTOR

Student Outcome

Students will compute the ratio of the area of the windows of a house to its square footage and express it as a percent.

Materials

• Window Percent worksheet
• Cabin 1 and Cabin 2

Time Required. 1 class period

Procedure

1. Tell students that the amount of glass on the exterior walls of a house affects its energy efficiency. Windows are a major source of heat loss in the winter and a large source of heat gain in the summer. A rule of thumb is that the area of the windows should be approximately 10% of the area of the house to provide a balance between natural lighting and excessive energy consumption. Thus if a home has 1,225 square feet of floor space, it should have about 122.5 square feet of windows.

2. Window sizes are shown on house plans. The number 4^03^6 beside a window means it is four feet, zero inches wide and three feet, six inches high. The first set of numbers always refers to width; the second refers to height. To find the area, multiply width by height.

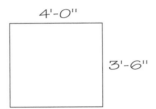

4 feet x $3\frac{1}{2}$ feet = 14 square feet

3. At first, students may find it confusing to convert 3 feet, 6 inches to $3\frac{1}{2}$ or 3.5.

4. Pass out the Window Percentage worksheet. Have students record the window sizes from Cabin 1 in the left column.

5. Have them compute the areas and find the total.

6. Next have them divide the total area of the windows by the square footage of the house. This should be rounded to two decimal places and then written as a percent.

7. Ask them if the result is approximately 10%.

Homework. Have students use the right-hand column of the Window Percent worksheet to analyze Cabin 2.

© Dale Seymour Publications®

NAME

WINDOW PERCENTAGE

List the windows of the house and find each area. Then divide the total by the square footage of the house. Write your answer as a percent.

Plan _____   Plan _____

Window size   Area   Window size   Area

1. _____ x _____   _____   1. _____ x _____   _____

2. _____ x _____   _____   2. _____ x _____   _____

3. _____ x _____   _____   3. _____ x _____   _____

4. _____ x _____   _____   4. _____ x _____   _____

5. _____ x _____   _____   5. _____ x _____   _____

6. _____ x _____   _____   6. _____ x _____   _____

7. _____ x _____   _____   7. _____ x _____   _____

8. _____ x _____   _____   8. _____ x _____   _____

9. _____ x _____   _____   9. _____ x _____   _____

10. _____ x _____   _____   10. _____ x _____   _____

11. _____ x _____   _____   11. _____ x _____   _____

12. _____ x _____   _____   12. _____ x _____   _____

13. _____ x _____   _____   13. _____ x _____   _____

14. _____ x _____   _____   14. _____ x _____   _____

Total window area   _____   Total window area   _____

Total house area   _____   Total house area   _____

Window area ÷ house area   _____   Window area ÷ house area   _____

(Round to two decimal places and write as a percent.)

© Dale Seymour Publications®

ELEVATION SORTING

Student Outcome

Students will match their floor plans to the correct exterior elevations. Students will also gain practice in computing the area of complex floor plans.

Materials

- partial sets of Plans 1–10 (See step 1 below to determine quantity.)
- complete class set of one plan for practice (Plan 1 works well for this.)
- elevations for houses 1–10 (See step 1 below to determine quantity.)
- rulers
- Architect's Record
- Elevation transparency

Time Required. 1–2 periods

Procedure

1. To reduce the amount of copying required to set up this lesson, you may wish to have each team work on a different house plan. Then you would only need one copy of a single plan for each member of one team. However, if you wish students to gain more practice, a full set of plans can be given to each student. In addition, instead of giving a complete set of elevations to every student, you can post one complete set on a bulletin board. Then students can match their floor plan to the correct elevations and remove them from the bulletin board if necessary.

2. Tell them they will need to study their plan and compute its square footage. To demonstrate this, have the students divide Plan 1 into shapes for which they can find the area. (Most students will have little trouble by now working with rectangles.) When they get to the bay window entry of Plan 1, they will probably come up with more than one way to find its area. Most will divide it into a rectangle and two triangles, which can then be formed into a square.

Step 1

Step 2

Step 3

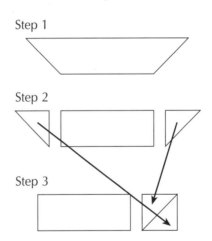

This is the opportunity to demonstrate formulas for the area of a triangle, parallelogram and trapezoid as shown on page 44.

The second form of the formula for a trapezoid (Area = median x height) is easier for middle school students to remember. It is easy for them to think of finding the number halfway between the two lengths of the bay. Then they simply multiply by how far the bay sticks out from the wall.

3. After the square footage is computed, students will need to find the four elevations that match their house. To demonstrate this, have them look at Plan 1. Study the placement of doors and windows. Try to picture how the roof would look on this house. Then examine the elevations. Which set would fit Plan 1? (The correct set are B, U, R, and Y.)

Note: For a more challenging lesson, each page of elevations can be cut into individual elevations, and mixed up.

4. Students can present their results orally to the class. Individual members can give different parts of the presentation. For example, one student can tell how the floor plan was divided into smaller shapes. Another student can explain how the square footage was computed. You may also require students to complete other activities from optional lessons such as Activity 6B if the concepts have been covered.

5. Extra plans have been provided in this unit. If students need extra practice or more challenge, they can be directed to Cabin 1, and Cabin 2, or full-size Plans A–D.

Extension. Students can create fairly accurate models by scaling down the floor of their plan to $\frac{1}{4}$ inch = 1 foot (half the normal scale) and taping the four elevations to its sides. A much more challenging approach is to scale up the elevations to the $\frac{1}{8}$-inch scale and tape it to the actual plan.

Journal Topic. (Display the Elevation transparency.) This elevation is a view of one of these two houses. Which house and which elevation is it, north, south, east, or west? Give as many reasons as you can to support your view.

Homework. Finish computing square footage on the other houses in the set. Match each plan to its elevations. Record your answers on the Architect's Record. (Have students use the median cost per square foot from Activity 4. This assignment may take students a few nights to complete.)

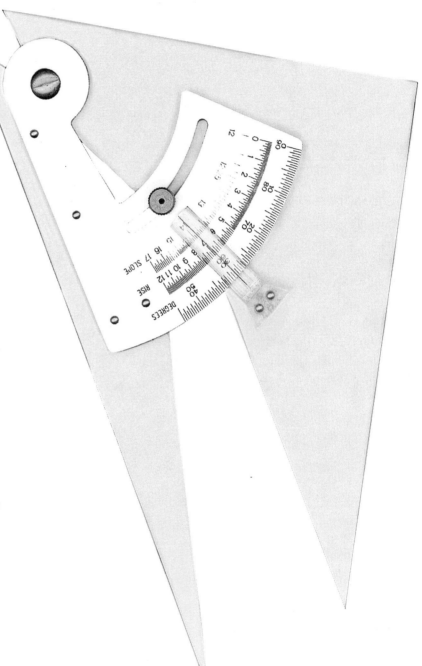

TRIANGLE

Area = $\frac{1}{2}$ base × height

PARALLELOGRAM

Area = base × height

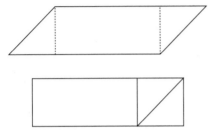

TRAPEZOID

Area = $\dfrac{\text{base}_1 + \text{base}_2}{2}$ × height

= median × height

Step 1

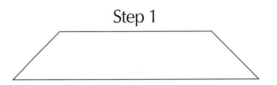

Step 2

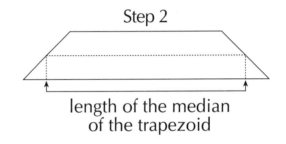

length of the median
of the trapezoid

Step 3

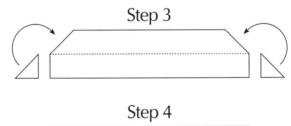

Step 4

© Dale Seymour Publications®

ARCHITECT'S RECORD

Cost per square foot _____

Plan Number	Square Footage	Total Cost	Elevations
1	_____	_____	_____
2	_____	_____	_____
3	_____	_____	_____
4	_____	_____	_____
5	_____	_____	_____
6	_____	_____	_____
7	_____	_____	_____
8	_____	_____	_____
9	_____	_____	_____
10	_____	_____	_____
Cabin 1	_____	_____	
Cabin 2	_____	_____	
A	_____	_____	
B	_____	_____	
C	_____	_____	
D	_____	_____	

© Dale Seymour Publications®

ARCHITECT'S RECORD KEY

Cost per square foot _____

Plan Number	Area Square Feet	Total Cost	Elevations	Volume Cubic Feet
1	980	_____	B, U, R, Y,	7810
2	1000	_____	W, A, N, D	7970
3	1260	_____	M, I, L, K	10110
4	1180	_____	J, E, T, S,	9440
5	1180	_____	C, H, O, P,	9440
6	1208	_____	BB, UU, RR, YY	9660
7	1310	_____	CC, HH, OO, PP	10500
8	1240	_____	MM, II, LL, KK	9920
9	1210	_____	JJ, EE, TT, SS	9700
10	1260	_____	WW, AA, NN, DD	10110
Cabin 1	730	_____		5820
Cabin 2	850	_____		6780
A	1210	_____	page 60	
B	1300	_____	page 61	
C	1470	_____	pages 62–63	
D	1360	_____	pages 64–65	

© Dale Seymour Publications®

ELEVATION

THE WILLIAMSTON

N
W   E
S

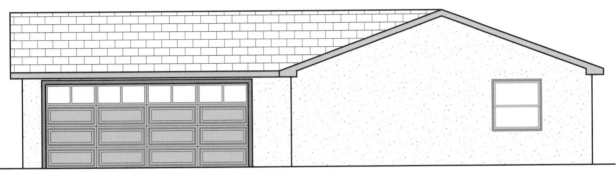

THE BRADFORD

N
W   E
S

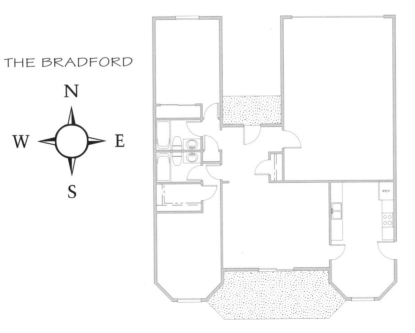

© Dale Seymour Publications®

ROOFING CONTRACTOR

Student Outcome

Students will compute the area of a roof, using the Pythagorean theorem.

Materials

- rulers
- Roof Diagrams 1–4
- transparency of Roof Diagrams 1

Optional

- Collect and display pictures of different style houses with different roof designs. Be sure to include ones with flat roofs as well as ones with steep pitches.

Time Required. 1 class period

Vocabulary

beams Heavy wooden framing members that support other parts of the structure.

ceiling The overhead inside lining of a room.

ceiling joists Wood framing members that support the ceiling.

gable roof A double-sloping roof with a single ridge in the middle and a gable (vertical triangle) at each end.

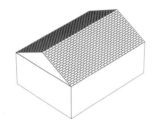

gambrel roof A curb roof with a lower steeper slope and an upper flatter one.

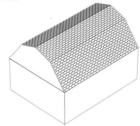

hip roof A roof with a single ridge, sloping ends, and sloping sides.

rafters Wood framing members that support the roof sheeting and roofing.

ridge The peak, or top, of the roof. The roof slopes down in both directions away from the ridge.

roof The cover of a building.

roof pitch or slope A measure of the rate at which the roof rises (rise divided by run).

roof truss A rigid framework made of several pieces of wood nailed or bolted together to form a combination roof and ceiling structure.

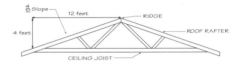

section A horizontal view of a house showing the construction features of what a house would look like if it were sliced by a huge knife.

shed roof A roof that has only one slope. The highest part of this type of roof is at one edge.

Procedure

1. Show students different examples of roofs. Explain that while a roof may have a pitch to it, the ceiling inside may be level at eight feet. Name the different roof styles and discuss the merits of them based on climate: steep roofs needed for snow climates, flat roofs good for hot, dry climates. The actual roof is usually layers of plywood covered with a moisture-proof layer, which is covered with the roofing material, either tiles, shingles, or gravel, depending on the style of the roof. Introduce the construction of trusses and rafters. Standard roof trusses come in 4:12 roof pitch.

2. Show students the overhead transparency of Roof Diagrams 1. Reveal only the top view. Tell them that this is the top view of a sloped roof. The building measures 24 feet wide and 20 feet long.

3. Ask them what the area of the roof is. Many students will say it is 480 square feet, although some students will see that this is not right.

4. Show them the lower part of the transparency. Show them that the roof is actually more than 24 feet wide. Ask them to estimate how wide each half is.

5. Pass out Roof Diagrams 1. Have students find the actual measurement of half the roof's width (13 feet).

6. First ask them if they know of a way to find that measurement without seeing all the measurements of the side view. Few if any students who have not used the Pythagorean theorem will know of a way. The Pythagorean theorem, $c^2 = a^2 + b^2$, allows builders to calculate the length of the hypotenuse, c, of a right triangle if they know the lengths of the two legs, a and b. This lesson provides practice using the Pythagorean theorem, but it is not intended to introduce or prove the formula.

7. Have students multiply the measurement of half the roof's width by the roof's length. This is the area of half the roof. Doubling it will give the total roof area.

8. Have them practice using the Pythagorean theorem by finding the area of the roofs of the two buildings on Roof Diagrams 2.

Extensions

• Ask students to compute the area of the roof of one of the house plans, assuming it has a 5:12 pitch. (That is, it rises 5 feet for every 12 feet of *run*, another name for horizontal distance.)

• Ask students to compute the cost of the roofing materials. Have them research costs of the materials common to your area and estimate costs based on two different materials choices. Tell them not to include the cost of the structure in their estimates, just the roofing layers.

Homework. Find the areas of the roofs of the buildings on Roof Diagrams 3 and 4.

ROOF DIAGRAMS 1

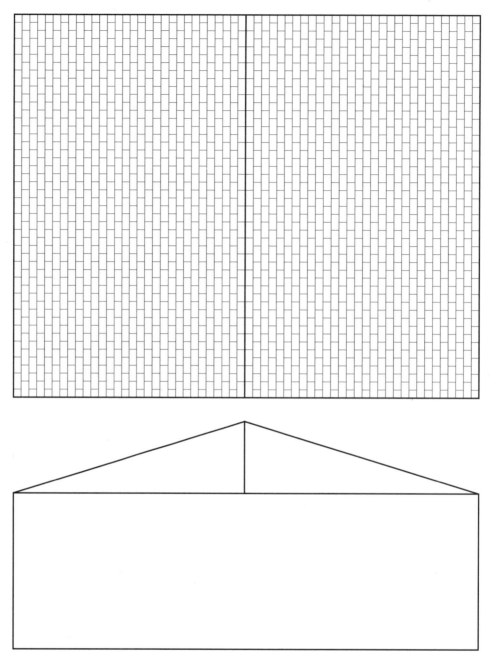

NOT TO SCALE

© Dale Seymour Publications®

NAME

ROOF DIAGRAMS 2

Use the Pythagorean theorem to find the missing values. Then find the area of the roof.

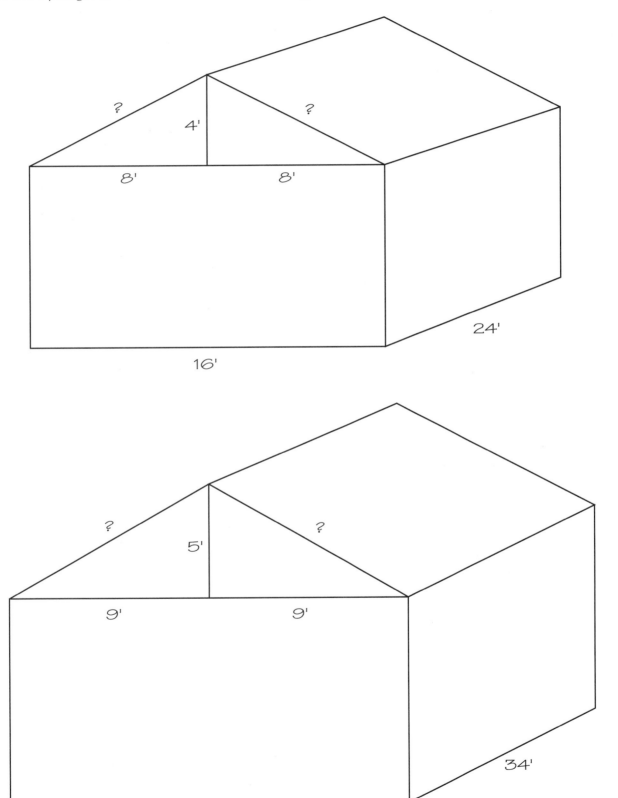

© Dale Seymour Publications®

ROOF DIAGRAMS 3

Use the Pythagorean theorem to find the missing values. Then find the area of the roof.

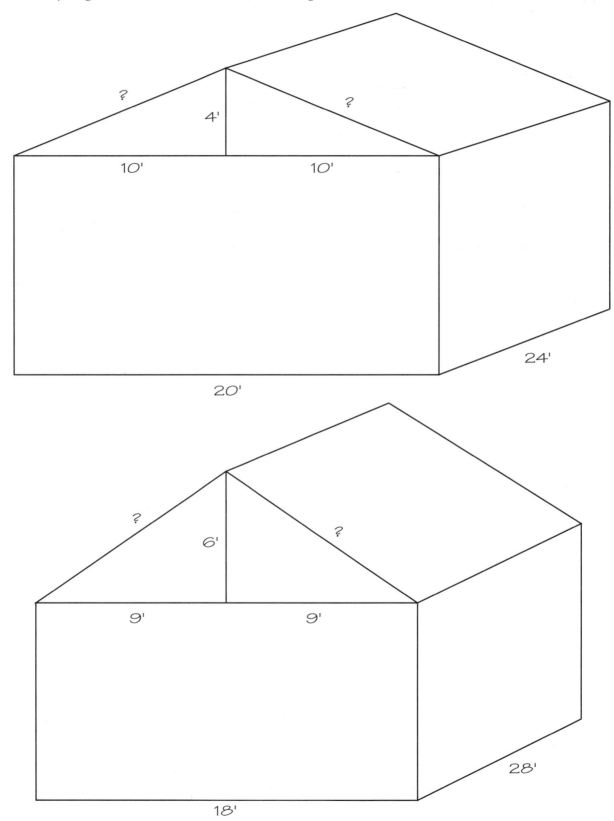

© Dale Seymour Publications®

ROOF DIAGRAMS 4

Use the Pythagorean theorem to find the missing values. Then find the area of the roof.

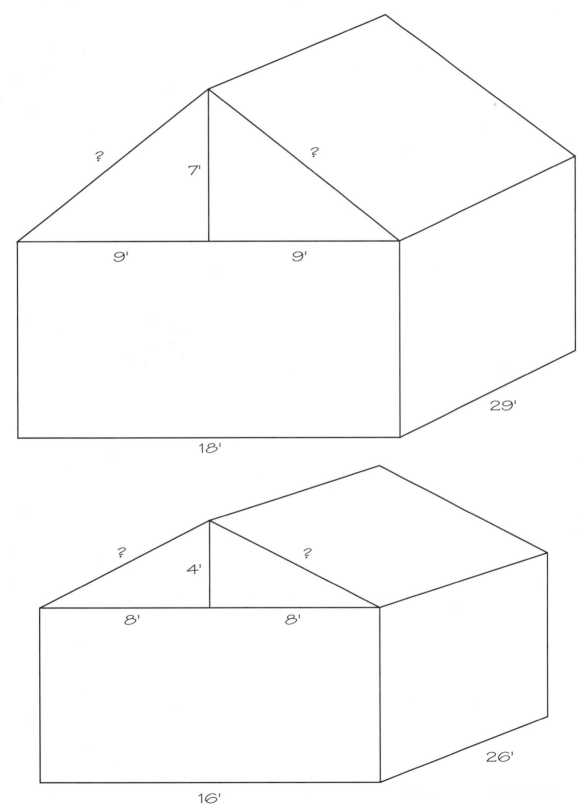

© Dale Seymour Publications®

DESIGNING A HOUSE

Student Outcome

Students will design a house within a given budget and compute square footage and cost of construction.

Materials

Required

- 12-inch-by-18-inch drawing paper
- grid paper
- Inspection Record (page 27)
- Requirements for Design
- corner triangles
- rulers
- Common Household Measurements

Optional

- drafting templates (You can purchase $\frac{1}{4}$-inch drafting templates. A class set may be expensive, but they last for years and give a professional touch to the house plans.)

Time Required. 2–5 periods

Procedure

1. Ask students to imagine they have been hired to design plans for three bedroom, two bath single-story homes. Each home must meet local building codes and developer guidelines, and must be under budget. Review and post the Requirements for Design. You may wish to embellish this by offering bonuses (extra credit) for creative and unique plans, or for plans allowing for wheelchair access.

2. Make available the plans that are included in this book for students to get ideas for their houses.

3. Have them use a pencil to lightly sketch their plan.

4. They are to then call the inspector (you) for a preliminary plan check. You may decide to let a fellow student do the preliminary plan check. This check will help students see what needs to be corrected, improve their final grade, and reduce the number of problems you need to check later.

5. Have them finalize their plan and complete the Inspection Record.

Extension. Students may be challenged by drawing some or all of the elevations for their house.

Journal Topics

- As an architect, write a note to a fellow worker, explaining how to find the area of the floor under a bay window.
- What have you learned during this housing unit that you did not know before? Give as many examples as you can. What are the three most important things you have learned?

Homework. Work on house plans. You may decide to have students do all work on house plans in class and continue the homework from Activity 7.

Assessment

If you have already done a preliminary plan check, it will be fairly easy to see if students have completed the assignment correctly. You can also reduce your workload by having a second student act as inspector. Tell students that when they finish their plan, they must get it checked by another student who signs as inspector.

You will probably find that students want to see every plan. You can accomplish this and get a head start on checking the plans by having students set their plans on their desks. Allow students to rotate throughout the class in groups of three or four. Let students study each set of plans for a few minutes. Then, before the next rotation, call on students randomly to mention something they liked about one of the plans in front of them. You can begin assessing some of the papers at this time.

REQUIREMENTS FOR DESIGN

- Each house must have at least three bedrooms, two full bathrooms, a full kitchen, a dining area, and a living room. There must be adequate closets and storage space for a family of four. Remember to include space for the hot water heater and the washer and dryer.

- The cost of the lot should not be a factor in the construction cost. The total construction cost for the house, including garage, must be under $200,000. The quality of construction to be used by the builder will cost $100 per square foot. Any roof style will be acceptable and not affect the cost.

- All houses should be designed to fit on a level building lot 80 feet wide by 112 feet deep. There must be side yards that are a minimum of 8 feet wide. The house must be set back from the front lot by 20 feet.

- You must use a standard two-car garage that measures 20 feet by 24 feet (480 square feet). The cost per square foot of the garage is half that of the rest of the house.

- Extra points will be awarded for houses that accommodate persons in wheelchairs.

- The project submission deadline is _____ . Projects turned in after this time will be reviewed but will automatically lose 20 points.

© Dale Seymour Publications®

ARCHITECTURAL SITE PLAN

Student Outcome

Students will draw a standard-size building lot and place the outline of the house drawn in Activity 8 on the lot, subject to developer guidelines.

Materials

- 12-inch-by-18-inch drawing paper
- Building Lot transparency
- drafting triangles
- rulers

Time Required. 1 period

Vocabulary

site plan A plan, usually drawn to $\frac{1}{8}$-inch scale ($\frac{1}{8}$ inch = 1 foot), that shows where a house is situated on the lot.

easement An area of the lot where utility companies can locate their service lines such as gas, electric, phone, cable, and water. Houses are not permitted to be built in easement areas.

property line A line that separates one building lot from another.

Procedure

1. Tell students that one of the requirements for getting a building permit is a site plan for their lot and house. For their site plan, the scale should be $\frac{1}{8}$ inch = 1 foot because they will have to show a larger area on their paper.

2. Show the Building Lot transparency to the class. All students will have the same size lot (in a subdivision) to build their home. Tell them that their houses may not be closer than 8 feet to a side property line, and may not be closer than 20 feet from the rear property line. No part of the house may lie in the easement area.

3. Ask what the maximum dimensions are that they can use in designing the house. How wide can the house be on the lot (64 feet)? How deep can the house be (72 feet)? Tell them that they must show the position of a garage and driveway so that the homeowner may add a garage at a later date. Note: If some students drew a house in Activity 8 that is too large to fit on the standard-size lot, suggest a modified house size that would fit on the lot.

Extension. Students can draw landscape details on the lot plan such as location of trees, shrubs, fences, and lawn.

Journal Topics

- Why is construction not allowed close to property lines or within easements?
- Why is it important to have a site plan?

BUILDING LOT

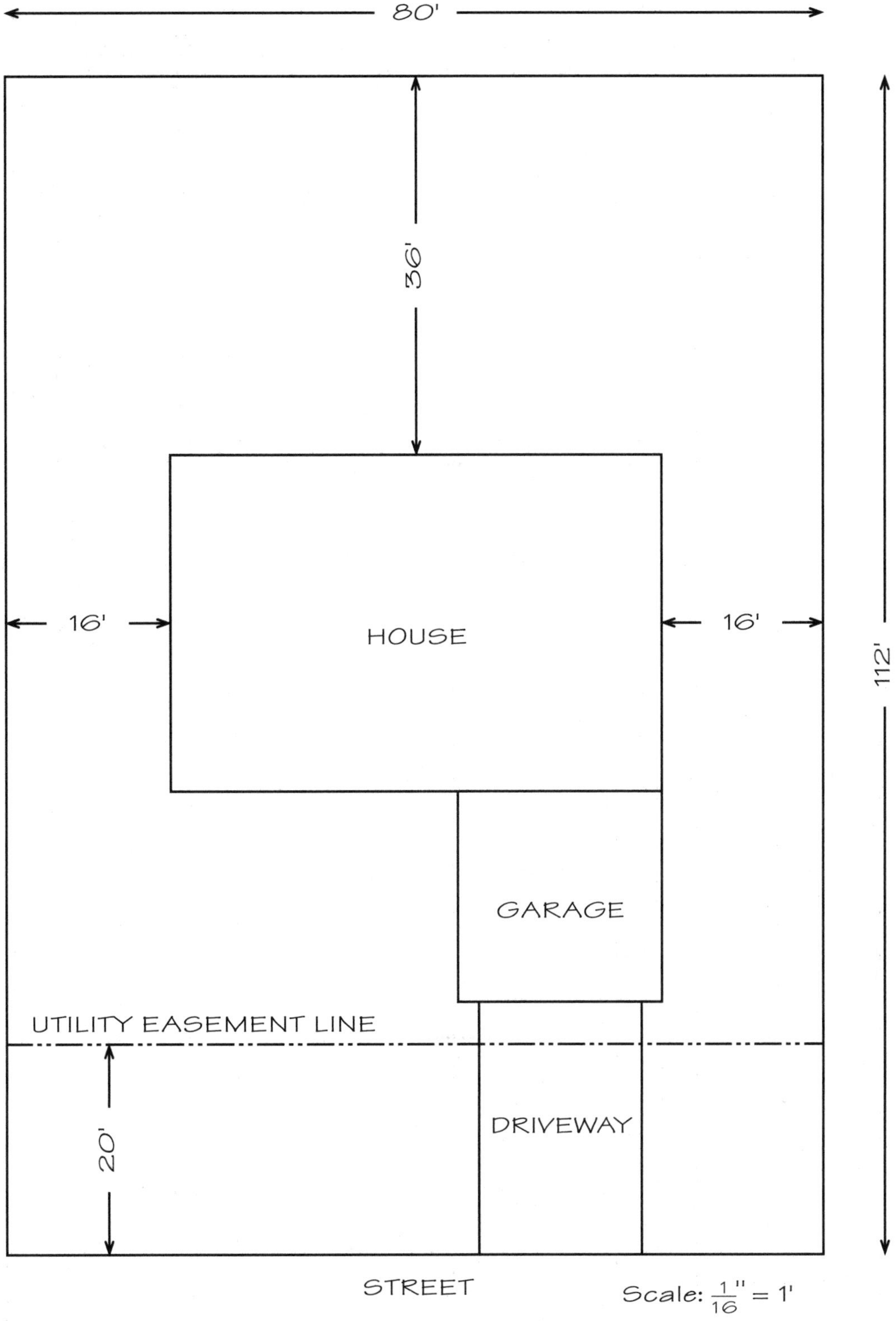

80'

36'

16'   HOUSE   16'

112'

GARAGE

UTILITY EASEMENT LINE

20'   DRIVEWAY

STREET   Scale: $\frac{1}{16}$" = 1'

© Dale Seymour Publications®

GLOSSARY

beams Heavy wooden framing members that support other parts of the structure.

building code Established by communities to ensure public safety in construction. Building codes ensure safe, sound, and sanitary buildings for people to live in.

easement An area of the lot where utility companies can locate their service lines such as gas, electric, phone, cable, and water. Houses are not permitted to be built in easement areas.

ceiling joists Wood framing members that support the ceiling.

electrical outlet Wall fixture into which electrical plugs are placed to get electricity; also called a *receptacle*.

elevation A view of a wall of a building as seen straight (perpendicular to the vertical surface). It illustrates the size, shape, and materials of the surface as well as the openings within it.

floor plan A view of a room or building looking straight down after the roof or upper floor is removed. It illustrates the size, form, and relationships between the spaces of the room measured horizontally, as well as the thickness and construction of the walls, columns, door and window openings, and fireplaces.

forced air heating A heater that circulates warm air through duct work.

furnace A heater that burns coal, oil, or wood to make heat.

gable roof A roof that has a single peak in the middle.

gambrel roof A curb roof with a lower steeper slope and an upper flatter one.

hip roof A roof with a single ridge, sloping ends, and sloping sides.

heat pump A heating engine that transfers heat from one area to another.

isometric drawing A view of a room or building as seen from above that represents the 3 dimensions of the space. Technically, the 90° corners of the room become 120° corners in the isometric drawing while all vertical elements remain vertical in the drawing. Isometric drawings can be quickly made on grid paper with horizontal lines drawn at a 30° angle from the horizontal.

linear foot One-foot measurement of an item along a line.

median In a set of data, the datum that is in the middle when numbers are ordered. Thus half the data values are lower than the median, and half are higher. The median home price is usually more indicative of typical local housing than the average. In the following example, the median home price is $128,500:
$115,000, $115,500
$128,500, $165,000
$279,000
In the next example, the median ($122,000) is the average of the two middle data:
$105,500, $115,000
$115,500, $128,500
$165,000, $279,000

property line A line that separates one building lot from another.

radiant heat A type of heat that uses radiation instead of forced air to heat a house.

rafters Wood framing members that support the roof sheeting and roofing.

ridge The peak, or top, of the roof. The roof slopes down in both directions away from the ridge.

roof pitch or slope A measure of the rate at which the roof rises (rise divided by run).

site plan A plan, usually drawn to $\frac{1}{8}$-inch scale ($\frac{1}{8}$ inch = 1 foot), that shows where a house is situated on the lot where it is to be built.

roof truss A rigid framework made of several pieces of wood nailed or bolted together to form a combination roof and ceiling structure.

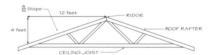

section A horizontal view of a house showing the construction features of what a house would look like if it were sliced by a huge knife.

shed roof A roof that has only one slope. The highest part of this type of roof is at one edge.

2 by 4 Standard size board. The actual size is $1\frac{1}{2}$ inches by $3\frac{1}{2}$ inches because lumber is measured to nominal size (2 inches by 4 inches) before being seasoned and planed.

© Dale Seymour Publications®

ARCHITECTURAL SYMBOLS

STANDARD ELECTRICAL OUTLET

220 VOLT ELECTRICAL OUTLET

LIGHT SWITCH

3-WAY LIGHT SWITCH

STANDARD CEILING LIGHT

FLUORESCENT CEILING LIGHT

CABINET WITH SINK

TOILET

BATHTUB

SHOWER

CEILING FAN/LIGHT

REFRIGERATOR

WASHER

DRYER

RANGE

DOUBLE-BASIN SINK

DISHWASHER

HOT WATER HEATER

WINDOW

SWINGING DOOR

SLIDING GLASS DOOR

SLIDING CLOSET DOOR

SHOWS FLOOR PLAN BELOW STAIR.
STAIR CONTINUES ABOVE CUT LINE.

© Dale Seymour Publications®

ELEVATION A

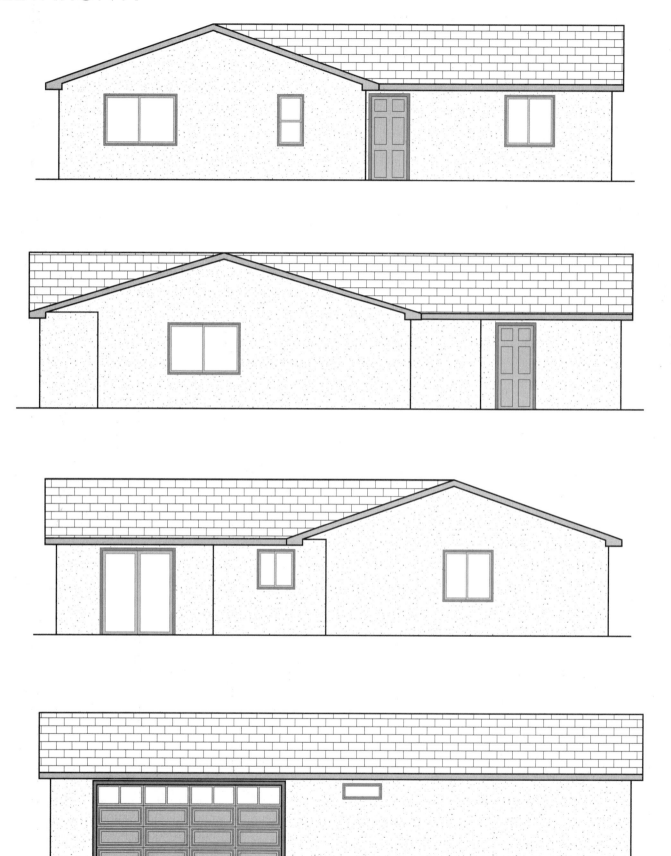

© Dale Seymour Publications®

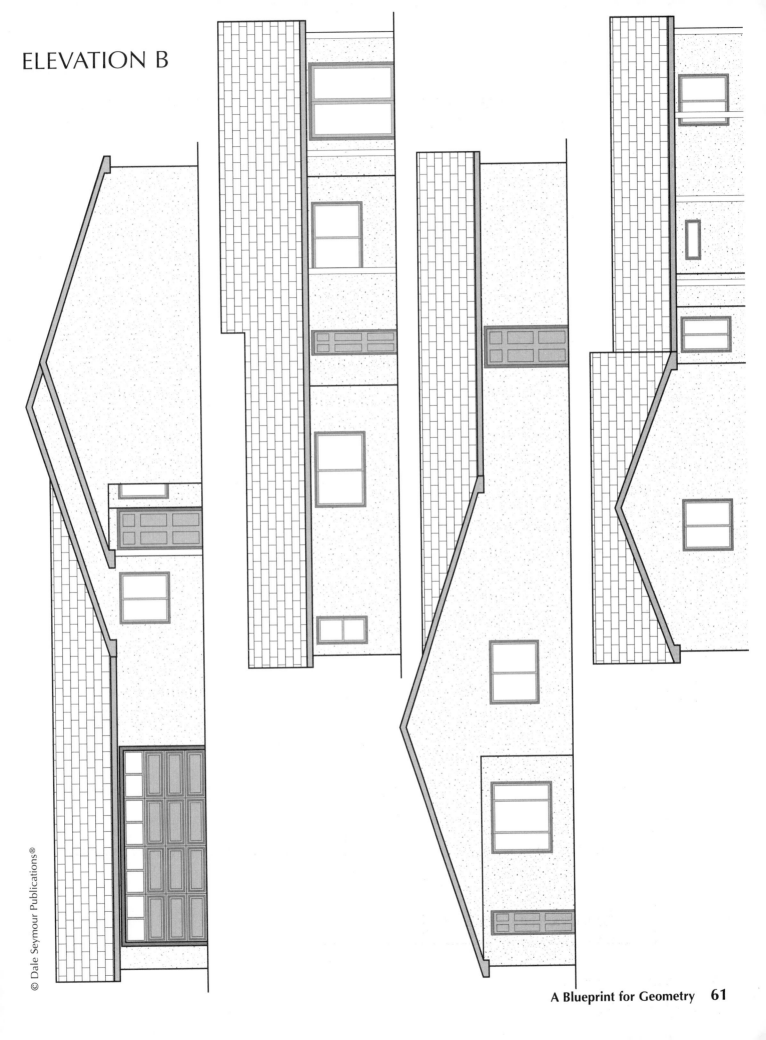

ELEVATION B

© Dale Seymour Publications®

A Blueprint for Geometry 61

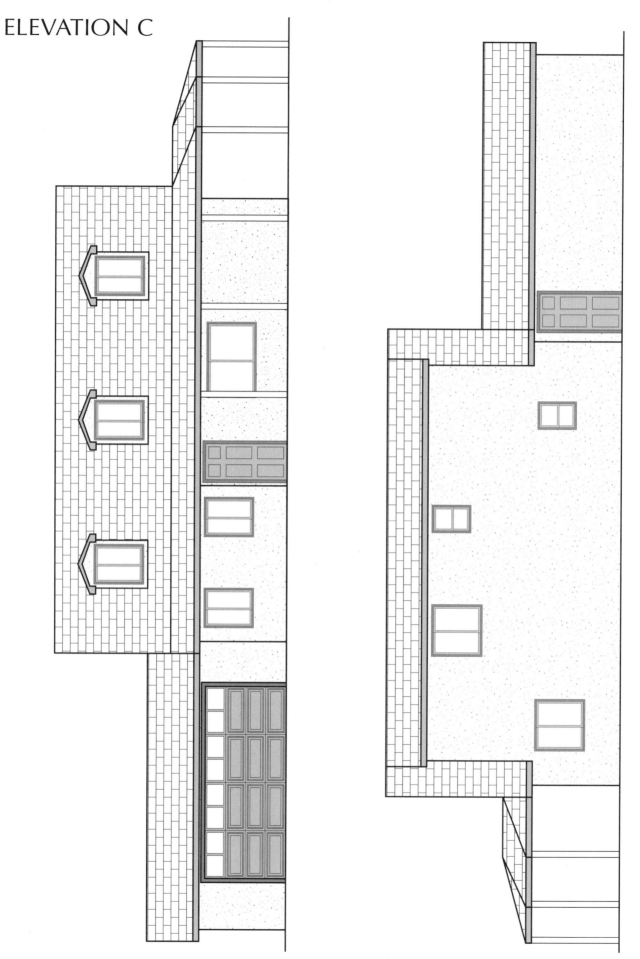

© Dale Seymour Publications®

ELEVATION C

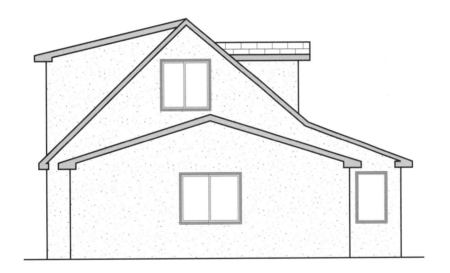

© Dale Seymour Publications®

ELEVATION D

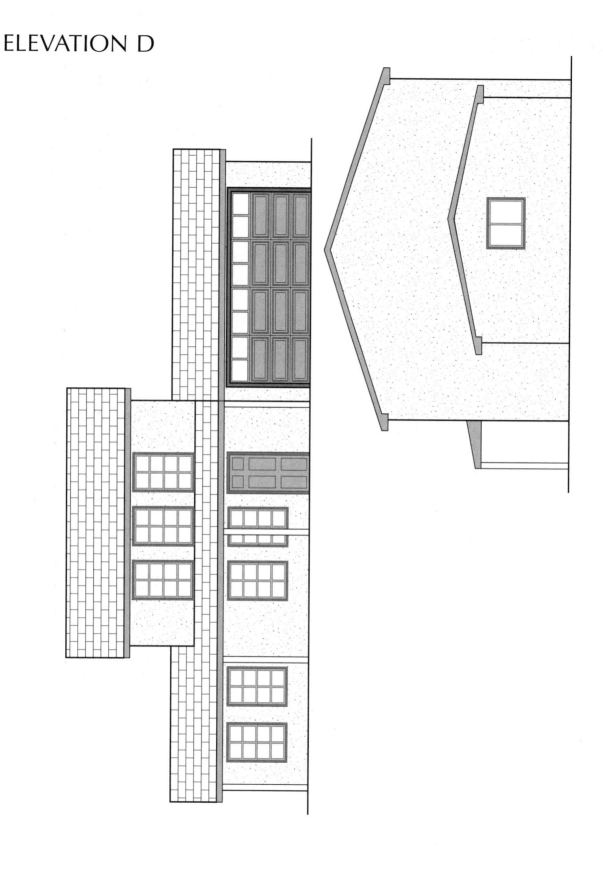

© Dale Seymour Publications®

ELEVATION D

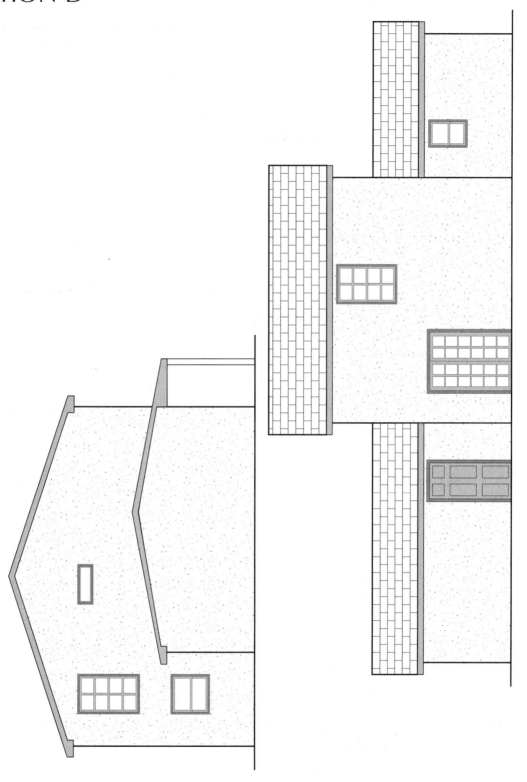

© Dale Seymour Publications®

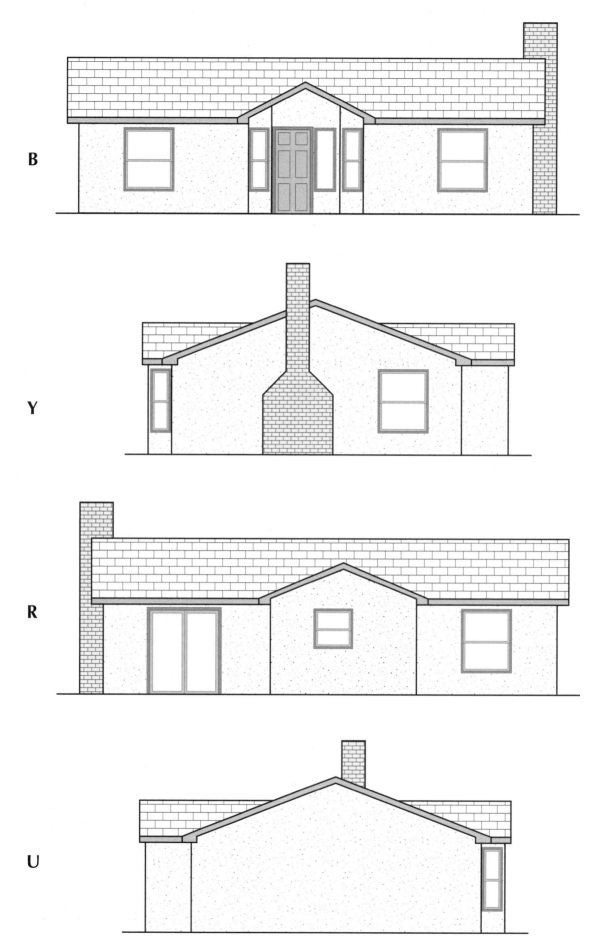

B

Y

R

U

© Dale Seymour Publications®

W

D

N

A

© Dale Seymour Publications®

A Blueprint for Geometry **67**

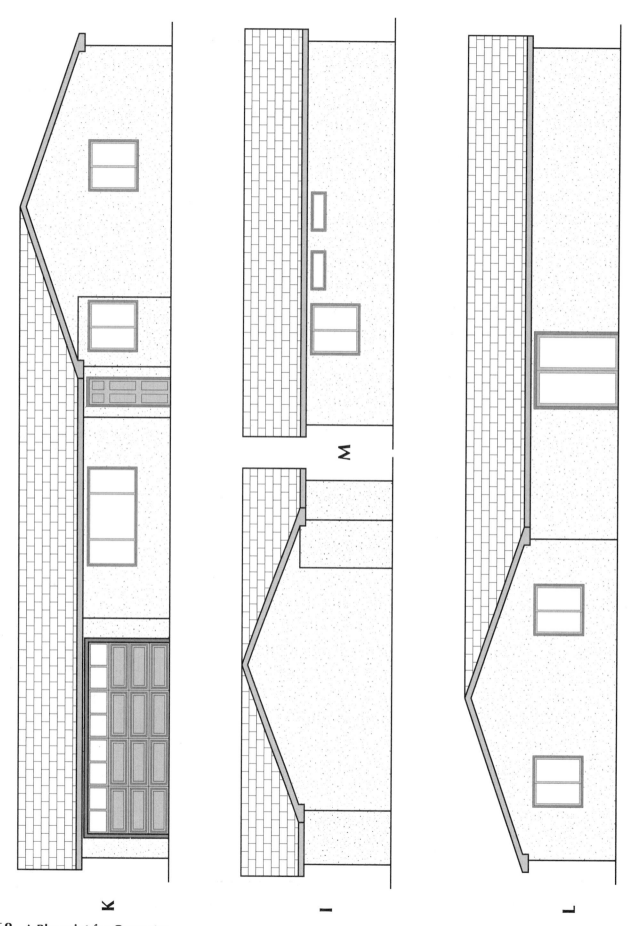

K

M

I

L

© Dale Seymour Publications®

© Dale Seymour Publications®

S

J

T

E

H

P

C

O

© Dale Seymour Publications®

BB

UU

YY

RR

© Dale Seymour Publications®

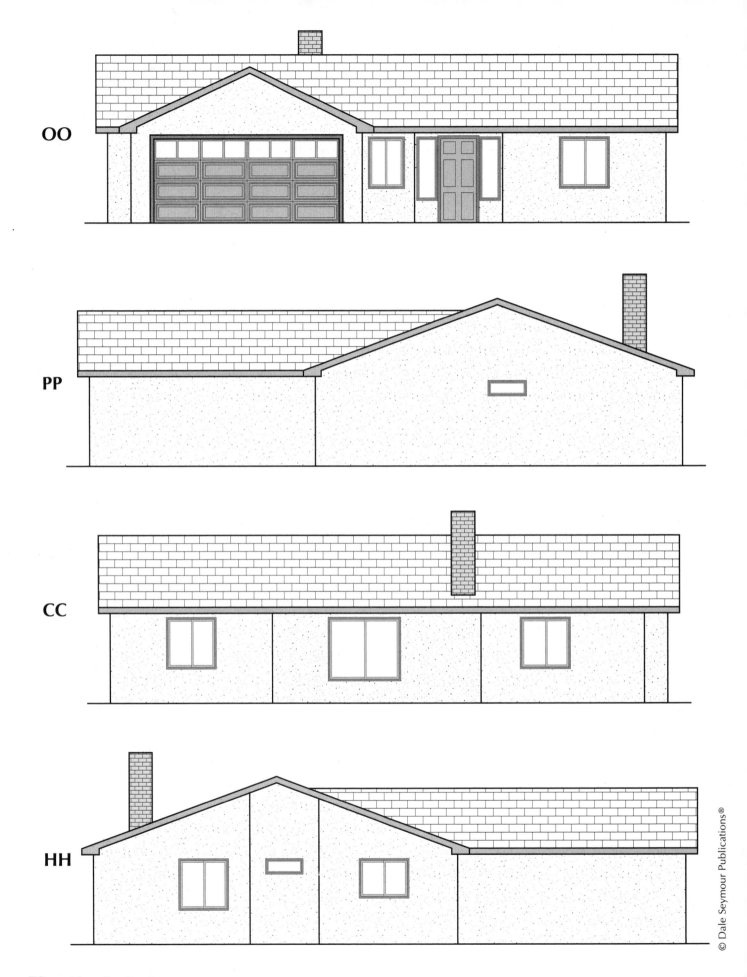

OO

PP

CC

HH

© Dale Seymour Publications®

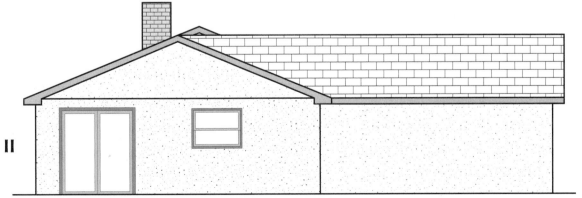

KK

MM

LL

II

© Dale Seymour Publications®

EE

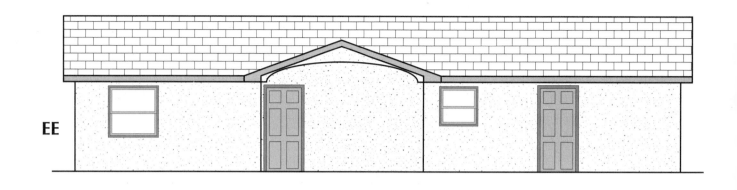

TT

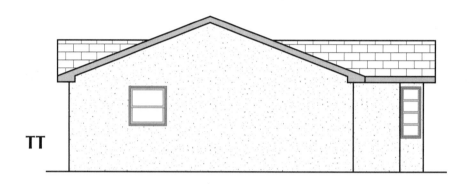

SS

JJ

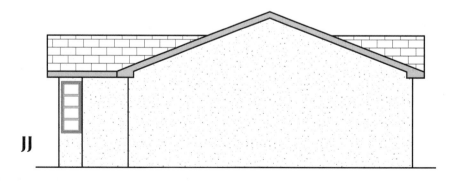

© Dale Seymour Publications®

WW

DD

NN

AA

© Dale Seymour Publications®

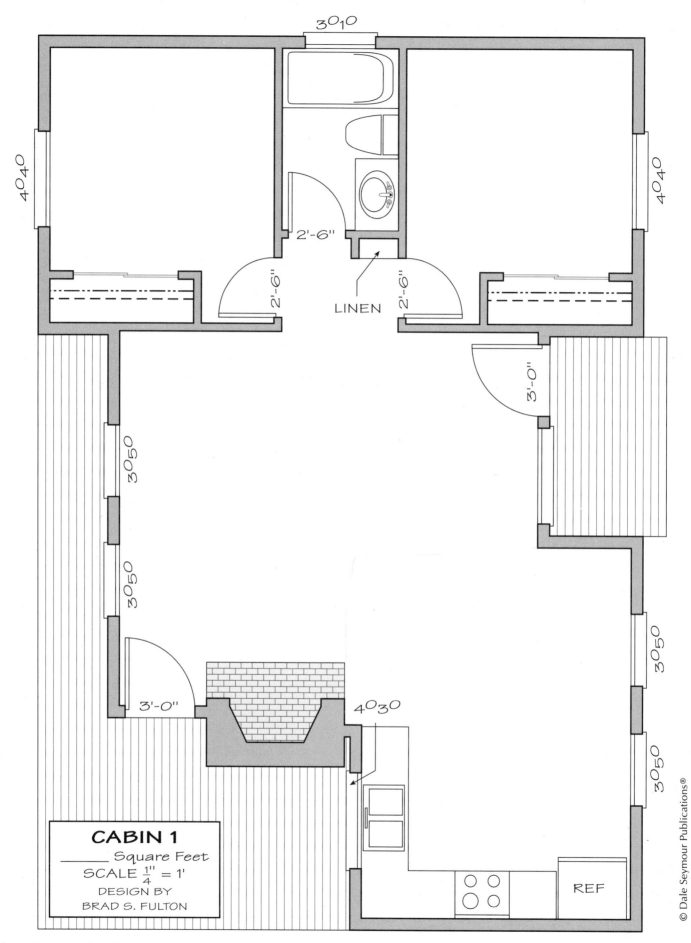

3010

4040

4040

2'-6"

2'-6"

LINEN

2'-6"

3'-0"

3050

3050

3050

3'-0"

3050

4030

REF

CABIN 1

_____ Square Feet

SCALE $\frac{1''}{4}$ = 1'

DESIGN BY
BRAD S. FULTON

© Dale Seymour Publications®

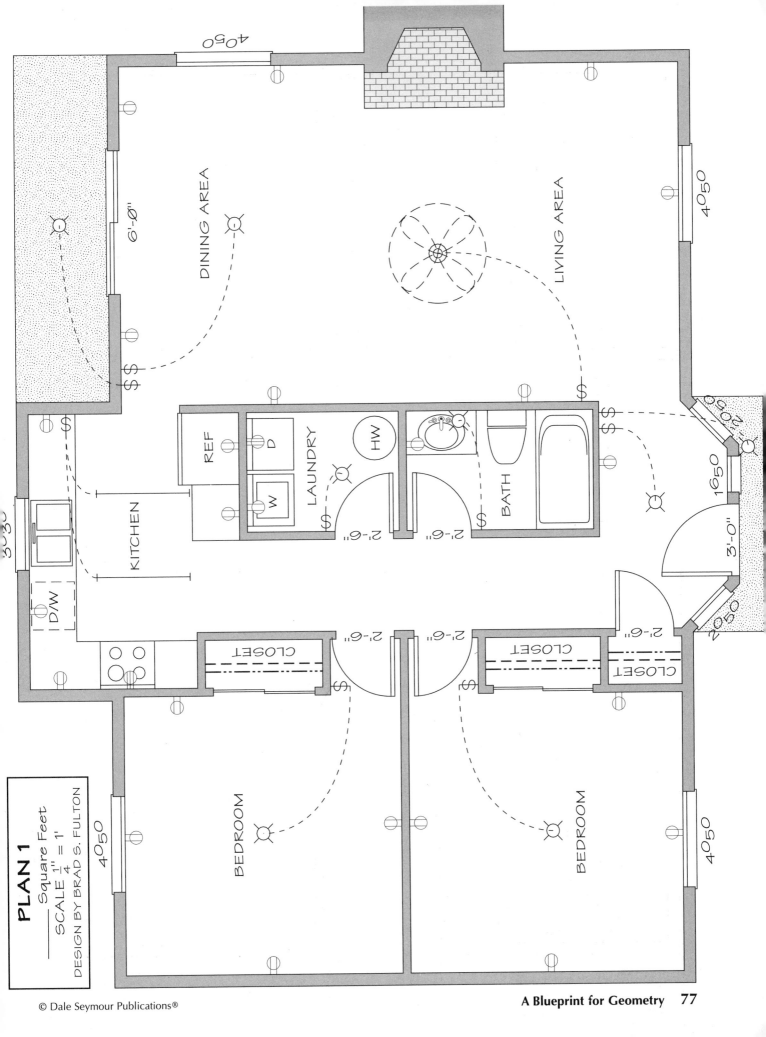

PLAN 1
Square Feet
SCALE $\frac{1}{4}$" = 1'
DESIGN BY BRAD S. FULTON

4050

DINING AREA

6'-0"

4050

LIVING AREA

REF

D

W

LAUNDRY

HW

BATH

KITCHEN

2'-6"

2'-6"

1650

3'-0"

2050

D/W

3030

CLOSET

2'-6"

2'-6"

CLOSET

CLOSET

2'-6"

BEDROOM

BEDROOM

4050

4050

© Dale Seymour Publications®

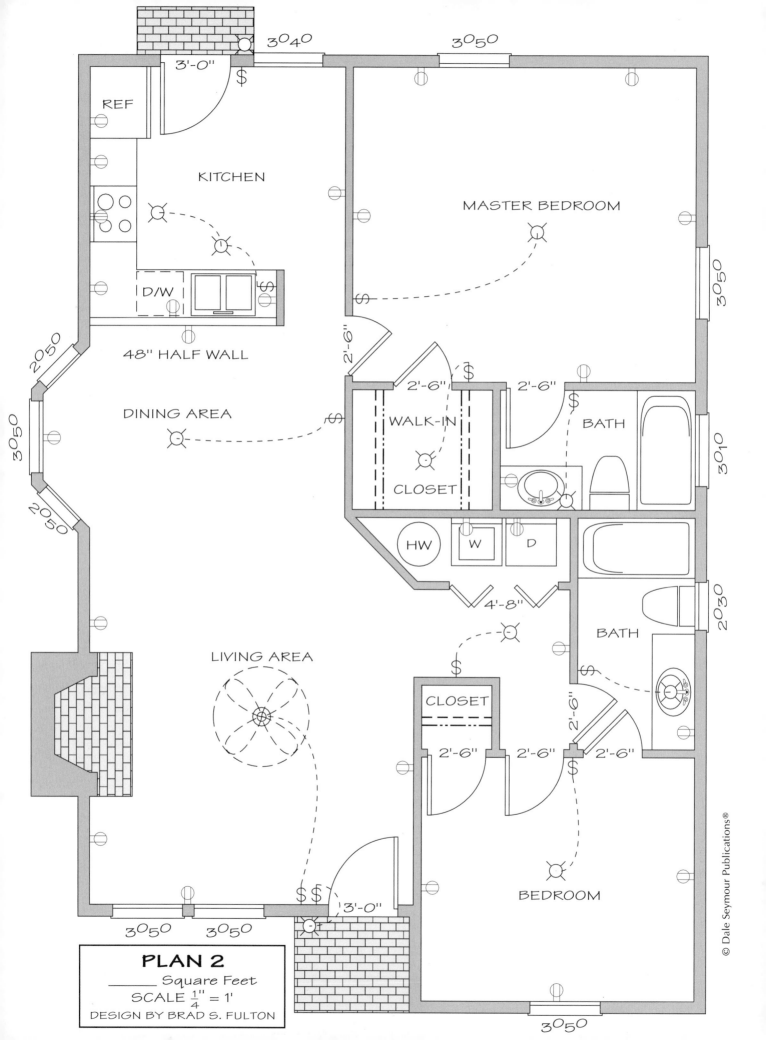

3040

3050

3'-0"

REF

KITCHEN

MASTER BEDROOM

3050

D/W

48" HALF WALL

DINING AREA

2'-6"

2'-6"

2'-6"

3050

WALK-IN

BATH

3010

CLOSET

2050

HW

W

D

BATH

LIVING AREA

4'-8"

CLOSET

2030

2'-6"

2'-6"

2'-6"

2'-6"

BEDROOM

3050

3050

3'-0"

3050

© Dale Seymour Publications®

PLAN 2

_____ Square Feet

SCALE $\frac{1}{4}$" = 1'

DESIGN BY BRAD S. FULTON

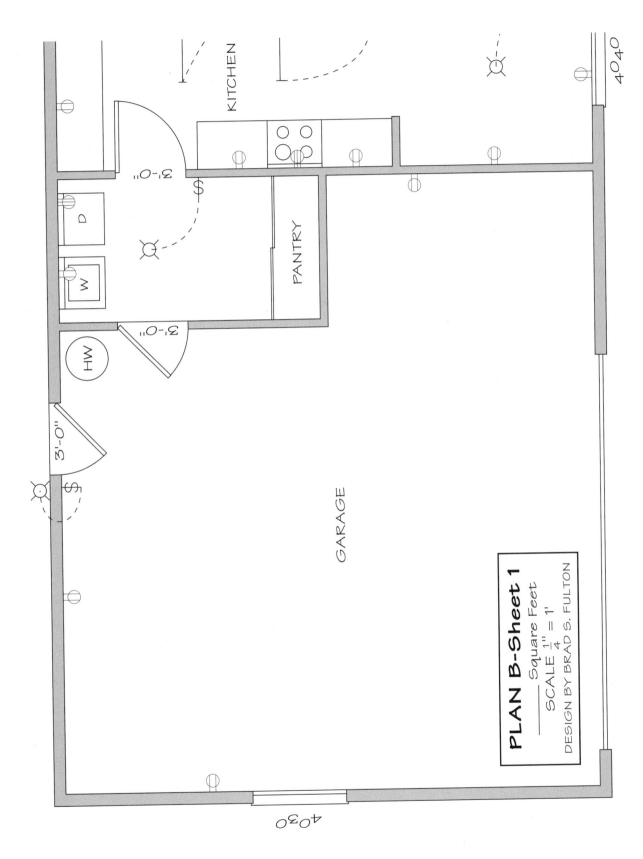

KITCHEN

4040

3'-0"

D

W

S

PANTRY

HW

3'-0"

3'-0"

S

GARAGE

PLAN B–Sheet 1

Square Feet

SCALE $\frac{1"}{4}$ = 1'

DESIGN BY BRAD S. FULTON

4030

© Dale Seymour Publications®

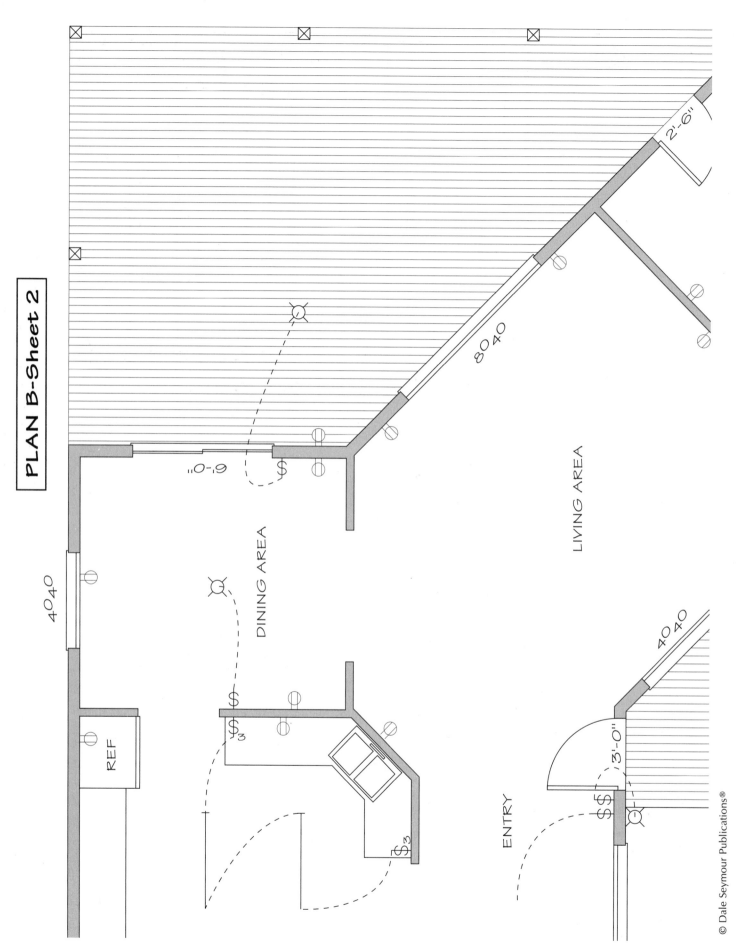

© Dale Seymour Publications®

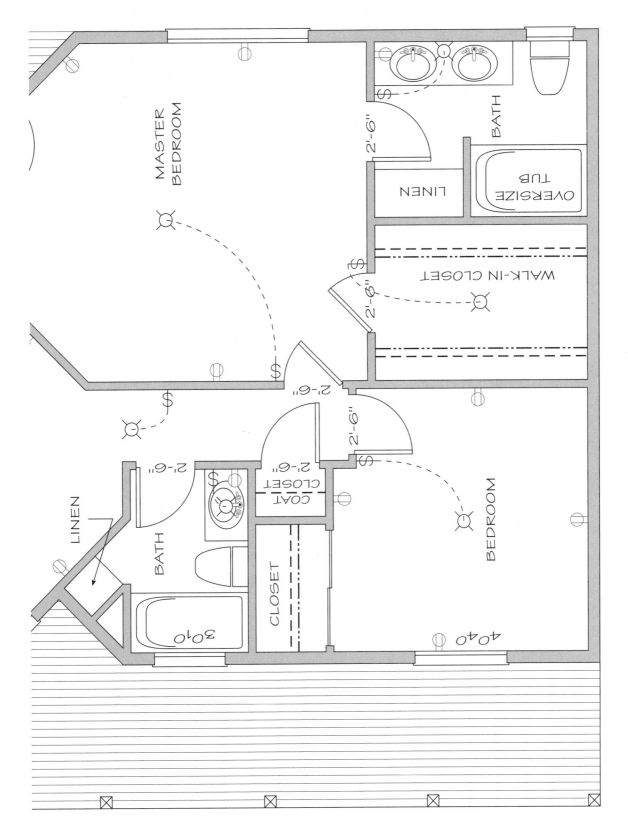

PLAN B-Sheet 3

MASTER BEDROOM

BATH

LINEN

OVERSIZE TUB

WALK-IN CLOSET

2'-6"

2'-6"

2'-6"

2'-6"

LINEN

BATH

3040

COAT CLOSET

CLOSET

2'-6"

BEDROOM

4040

© Dale Seymour Publications®

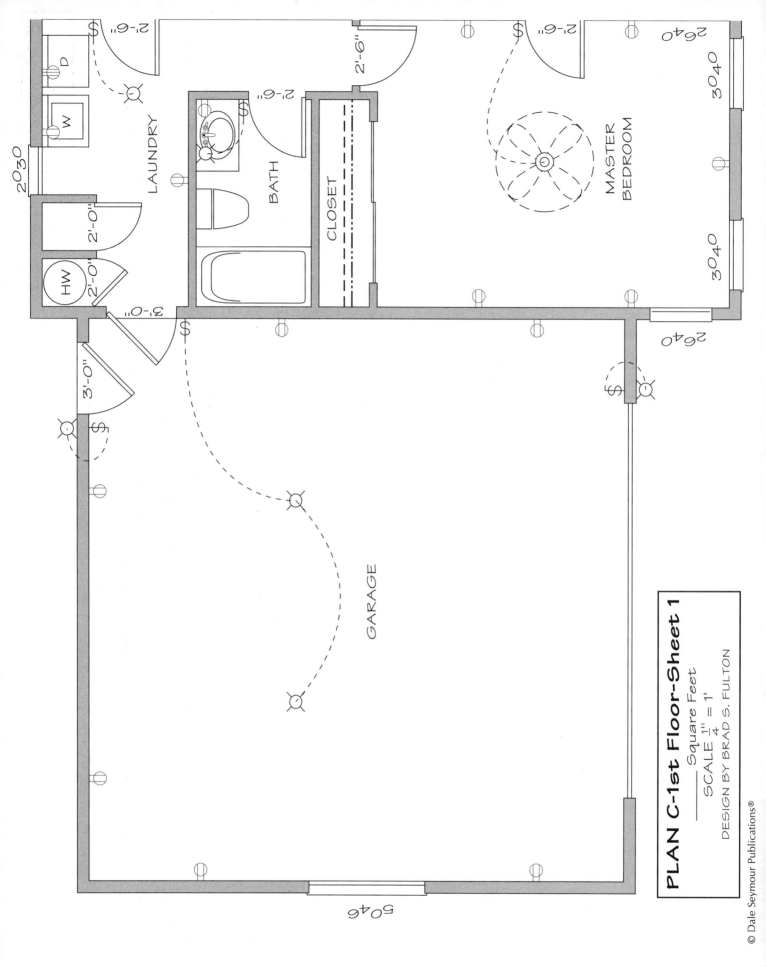

PLAN C-1st Floor-Sheet 1

_____ Square Feet

SCALE $\frac{1}{4}$" = 1'

DESIGN BY BRAD S. FULTON

LAUNDRY

BATH

CLOSET

MASTER BEDROOM

GARAGE

W

D

HW

2-6"

2-6"

2-6"

2-6"

2-0"

2-0"

3-0"

3-0"

2030

2640

3040

3040

3040

2640

5046

© Dale Seymour Publications®

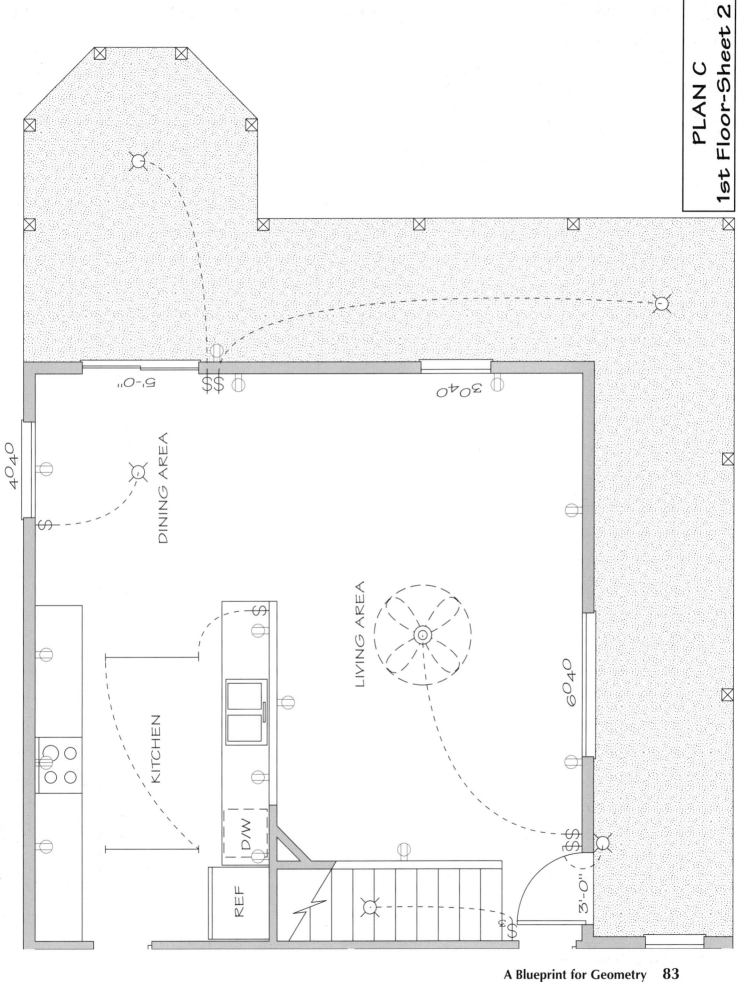

DINING AREA

KITCHEN

LIVING AREA

REF

D/W

5'-0"

4040

3040

6040

3'-0"

© Dale Seymour Publications®

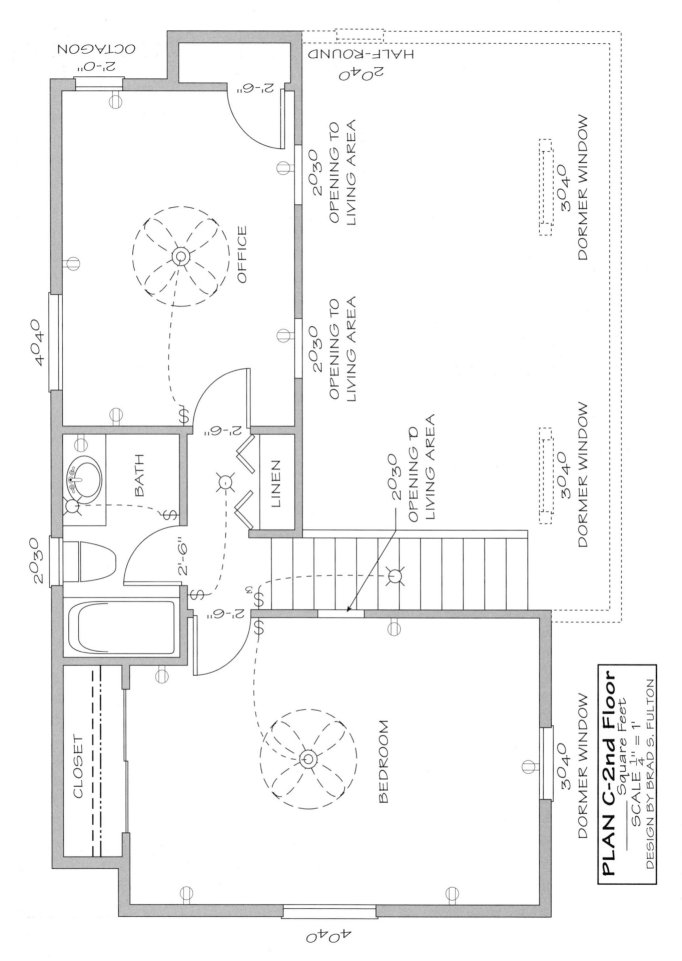

PLAN C-2nd Floor
Square Feet
SCALE $\frac{1}{4}'' = 1'$
DESIGN BY BRAD S. FULTON

OCTAGON
2'-0"

HALF-ROUND
2040

2040

3040
DORMER WINDOW

3040
DORMER WINDOW

2030
OPENING TO LIVING AREA

2030
OPENING TO LIVING AREA

2030
OPENING TO LIVING AREA

4040

OFFICE

2'-6"

BATH

2030

2'-6"

2'-6"

LINEN

2'-6"

CLOSET

BEDROOM

DORMER WINDOW
3040

4040

© Dale Seymour Publications®

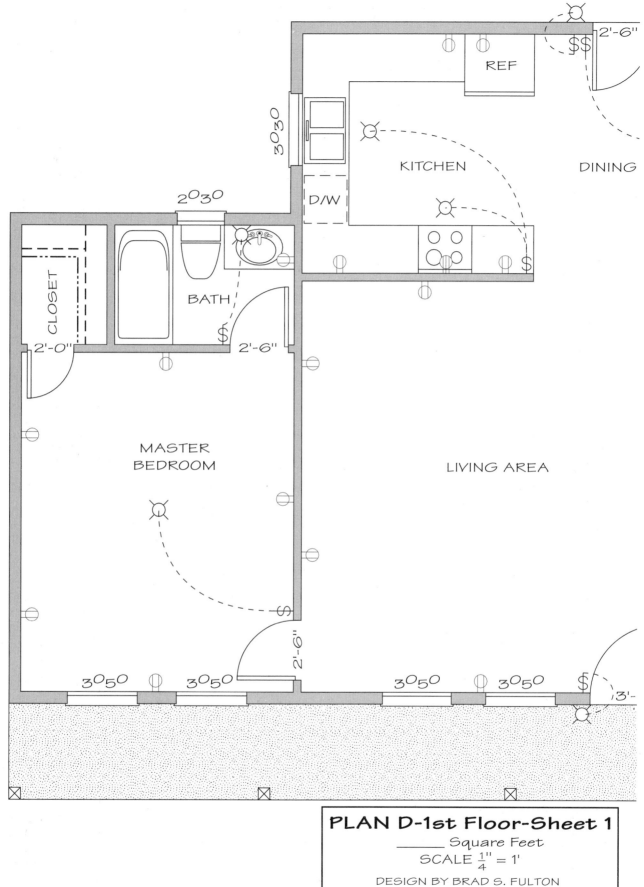

2'-6"

3030

REF

KITCHEN

DINING

2030

D/W

CLOSET

BATH

2'-0"

2'-6"

MASTER
BEDROOM

LIVING AREA

3050   3050

3050   3050

2'-6"

3'-

© Dale Seymour Publications®

PLAN D-1st Floor-Sheet 1

_____ Square Feet

SCALE $\frac{1''}{4} = 1'$

DESIGN BY BRAD S. FULTON

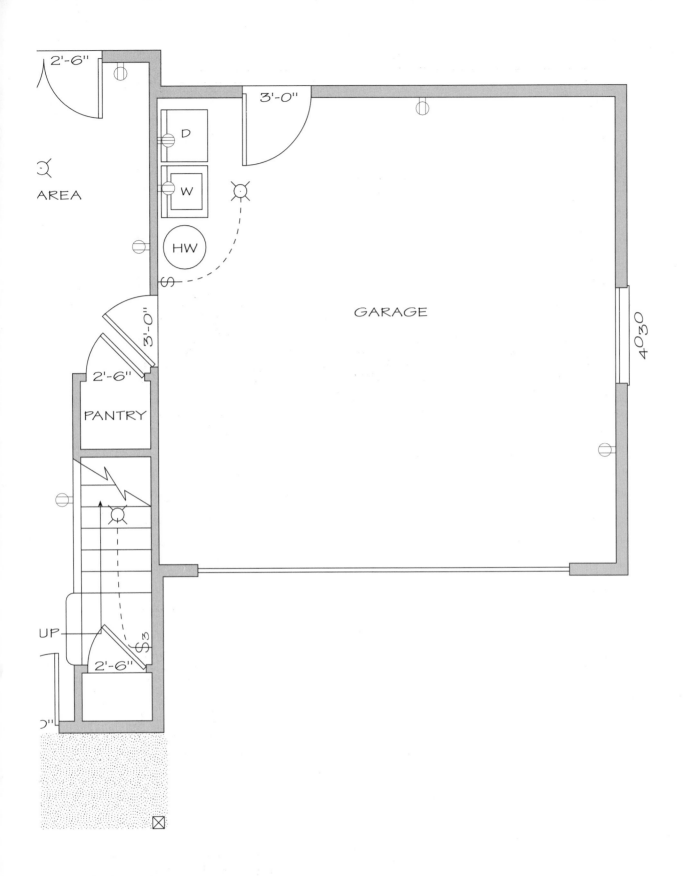

2'-6"

AREA

3'-0"

D

W

HW

S

3'-0"

2'-6"

PANTRY

4030

UP

S₃

2'-6"

0"

GARAGE

© Dale Seymour Publications®

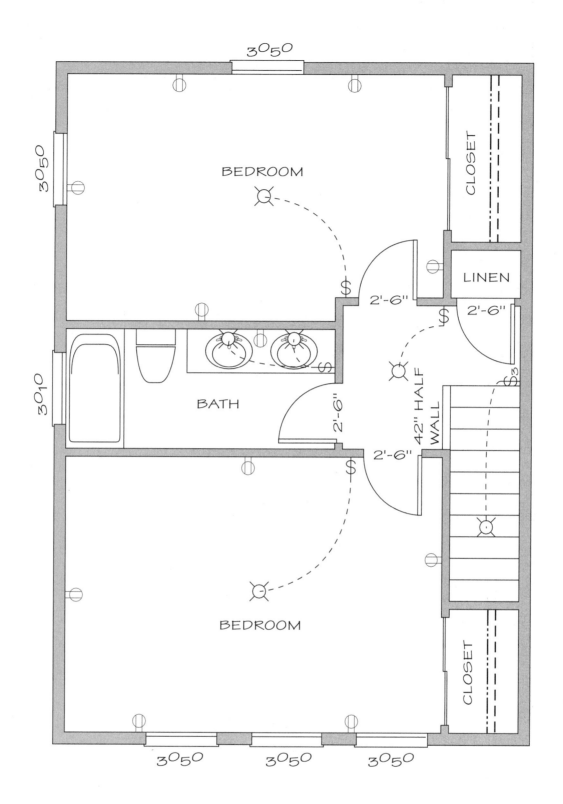

3050

3050

BEDROOM

CLOSET

LINEN

2'-6"

2'-6"

3010

BATH

2'-6"

42" HALF WALL

2'-6"

BEDROOM

CLOSET

3050   3050   3050

© Dale Seymour Publications®

PLAN D-2nd Floor
_____ Square Feet
SCALE $\frac{1''}{4} = 1'$
DESIGN BY BRAD S. FULTON

A Blueprint for Geometry   87

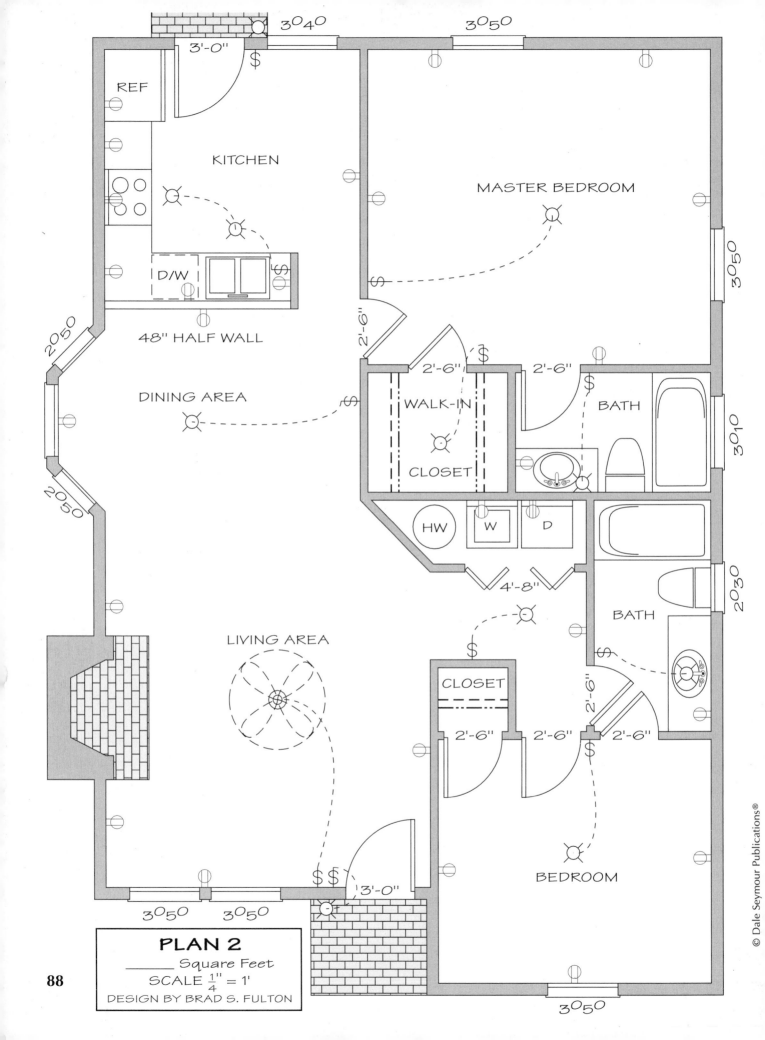

3^040

3^050

3'-0"

REF

KITCHEN

MASTER BEDROOM

3^050

2^050

D/W

48" HALF WALL

2'-6"

WALK-IN

2'-6"

BATH

3^010

DINING AREA

CLOSET

2^050

HW   W   D

4'-8"

BATH

2^030

LIVING AREA

CLOSET

2'-6"

2'-6"

2'-6"

2'-6"

3'-0"

BEDROOM

3^050 3^050

PLAN 2
_____ Square Feet
SCALE $\frac{1"}{4}$ = 1'
DESIGN BY BRAD S. FULTON

3^050

© Dale Seymour Publications®